摩登样板间 II
Modern Sample Houses II

欧式新古典
European Neo-Classical Style

ID Book 工作室 编

華中科技大學出版社
http://www.hustp.com
中国 · 武汉

The Beauty of Neo-Classicism

For quite a long time, people have never halted their pursuit of art. In the art movement emerging one after another, under the guidance of common tenets and purpose, the artists affect the developing orientation of world culture with the posture of avant-guard thinking. The Neo-Classicism has profound influence till today. It is dated back to the mid of 18th century, when people were against the complicated decorative style of Rococo and aspired to go back to the age with pure art of ancient Greece and ancient Rome.

Neo-Classical style influences a lot of areas such as decorative art, architecture, painting, literature, etc. After several centuries, this style is still as dazzling and magnificent as pearl. After being in the industry of interior design for more than 10 years, I am profoundly motivated by the arousing of inspirations by this style and the everlasting mastering of beauty. The design essence of loose form but condensed spirit aroused the emotions from the bottom of people's heart. During the design process, I have always been pursuing perfect graphic composition and making use of magnificent, elegant and epic like visual effects to move people's heart. This pure attitude in the creation is integrated with the honing of life, which is finally represented in the work itself.

With the current social background, the beauty of traditional spiritual temperament is retained and there appears the visual impression of extreme novelty. What is more, Neo-Classicism is retained as some renewed beauty which plays some irreplaceable function in the art current, producing some abundant and various creative languages for modern design field.

Variations with positive significance is always comforting. We are quite satisfied with the fact that there is the trend for Neo-Classicism in the time. A good design must can add some personal creative ideas for the more profound aesthetic thinking. And that is where the quality of Neo-Classicism lies.

Liu Weijun
PINKI Interior Design & IARI Interior Design Co., Ltd.

前言

新古典主义之美

长久以来，人类对于艺术的追求一直没有停止过。层出不穷的艺术运动中，艺术家在共同的宗旨和目标指引下，以思想先锋的姿态影响着世界文化的走向。至今仍具有深刻影响的新古典主义始于18世纪中叶，反映了人们对洛可可繁复装饰风格的反对，渴望回到古希腊和古罗马时期艺术的“纯洁”。

新古典主义风格影响了装饰艺术、建筑、绘画、文学等众多领域，在经历几个世纪以后，依然如珍珠般华彩万千。在室内设计行业从业十余年来，我对新古典主义带给我灵感的催生和对恒久美的把握深有感怀。其“形散神聚”的设计精髓能够唤起人们内心深处的情感。我在设计过程中，也一直在追求完美的构图，运用大气、典雅、史诗般的视觉效果打动人的内心。这种创作上的纯真态度，融合了生活的磨砺迸发出来，并最终在作品中得到展现。

在现在的社会背景下，具有传统精神气质的美被保留下来，并添加进了新颖、精致的视觉感受，新古典主义被以一种“更新”之后的美而保存下来，在艺术洪流里发挥着不可替代的作用，为当代的设计领域带来丰富而多样的创作语言。

有正面意义的变化永远是令人欣慰的，我们很欣慰这个时代依然涌动着新古典主义的激浪。好的设计，一定要在深层次的美学思想上增添个性的创意，这便是新古典主义的可贵之处。

刘卫军

PINKI品伊创意集团&美国IARI刘卫军设计师事务所

CONTENTS

目录

传麒山12栋B3户型样板房

Chuanqishan Show Flat, Building No.12, Flat Type B3

设计单位：戴维斯室内装饰设计（深圳）有限公司
设 计 师：Thomas（HK）、Marco
项目面积：133 m^2
主要材料：大理石、玉石、陶瓷锦砖、扪布、扪皮、壁纸、黑钢、精钢、玫瑰钢、特色玻璃、银镜、木饰面板、地毯

Design Company: Davis Interior Design (Shenzhen)Co., Ltd.
Designers: Thomas(HK), Marco
Project Area: 133 m^2
Major Materials: Marble, Jade, Ceramic Mosaic Tile, Leather, Wallpaper, Black Stell, Rose Steel, Special Glass, Silver Mirror, Wood Veneer, Carpet

本案运用现代法式的设计手法，以古典浪漫主义作为整体空间氛围的基调，采用新颖的材质语言来表达法式的浪漫。整体空间以淡灰色调为主，运用金色作为点缀，彰显出奢华的风尚。

从入户门进入，首先映入眼帘的就是一个特色门厅。独特的艺术玻璃配合个性的设计手法，再加上优雅、简洁的家具，给人留下深刻的印象。一块典雅的地毯将客厅与餐厅连接起来。客厅墙面的装饰采用欧式古典纹样，低调中流露出唯美的气息。顶棚饰线上扭花造型的法式线条、墙面上的特色壁纸和特色油漆搭配着曲线优美的家具，使整个空间充满了高雅的气息。

主卧背景墙上的特色壁纸、经过特殊处理的玻璃门、高贵典雅的水晶吊灯，呈现出了别样的温馨。亮丽的柠檬黄色床头背景与充满趣味的拼接图案相互融合，为空间注入一抹清新、自然的动感。卫生间以灰白色为主，简约的造型元素增强了空间感。

This project applies modern design approach and displays French romance with novel materials language with classical romanticism as the tone for the whole space atmosphere. The whole space is focused on light gray color tone and applies gold as the ornaments, displaying the fashion of luxury.

Upon entering the entrance, what comes into eyes first is a special entrance hall. The peculiar artistic glass is accommodated with personal design approach, added with elegant and concise furniture, which gives people some profound impressions. An elegant carpet connects the living room with the dining hall. The decoration of the living room's wall uses European classical patterns, presenting some aesthetic atmosphere in the low-key style. The French lining on the ceiling, the special wallpaper on the wall, the special paint and the furniture with delicate curves make the whole space full of elegant atmosphere.

The special wallpaper on the master bedroom's background wall, the glass door after special treatment, the elegant crystal chandeliers display some peculiar warmth. The light lemon bedside background integrates with the collage graphics of great fun instilling some fresh and natural dynamic feelings. The washroom is focused on gray white color tone. The concise format elements enhance the space feel.

Robinson Crusoe
Robinson Crusoe

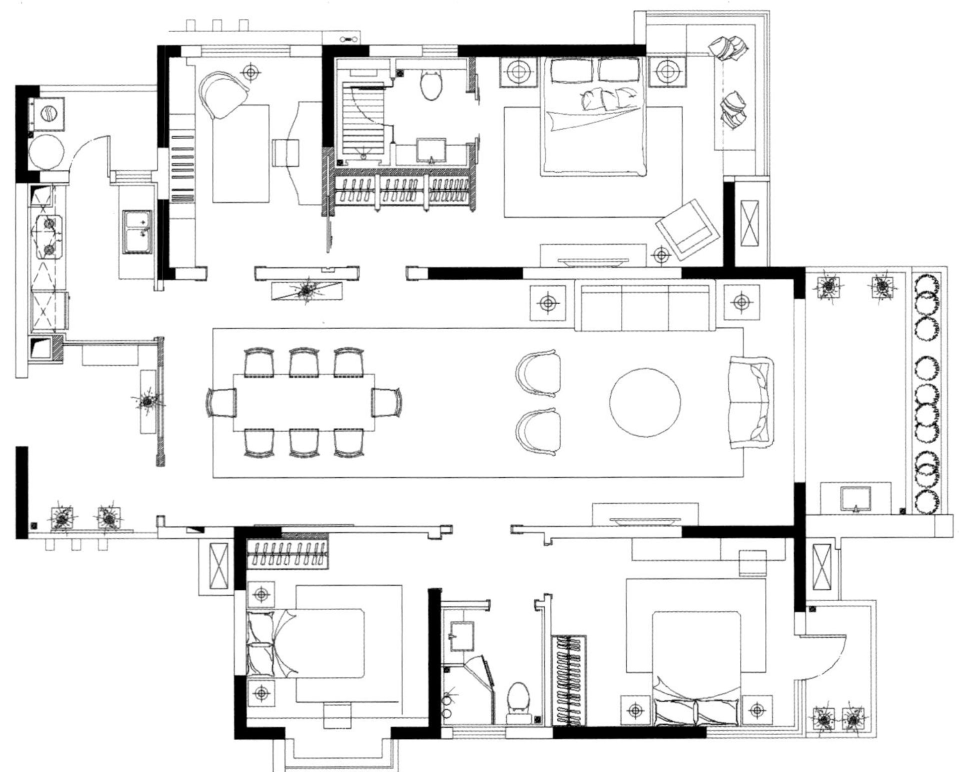

宝安天御样板间

Bao'an Tianyu Show Flat

设计单位：深圳市盘石室内设计有限公司	Design Company: Shenzhen Panshi Interior Design Co., Ltd.
设 计 师：吴文粒	Designer: Wu Wenli
项目地点：广东省深圳市宝安区	Project Location: Bao'an District, Shenzhen in Guangdong Province
项目面积：167 m^2	Project Area: 167 m^2
主要材料：意大利灰木纹大理石、皮革、黑檀木、亮面玫瑰金	Major Materials: Italian Gray Wood Grain Marble, Leather, Black Teak, Light Rose Gold Marble

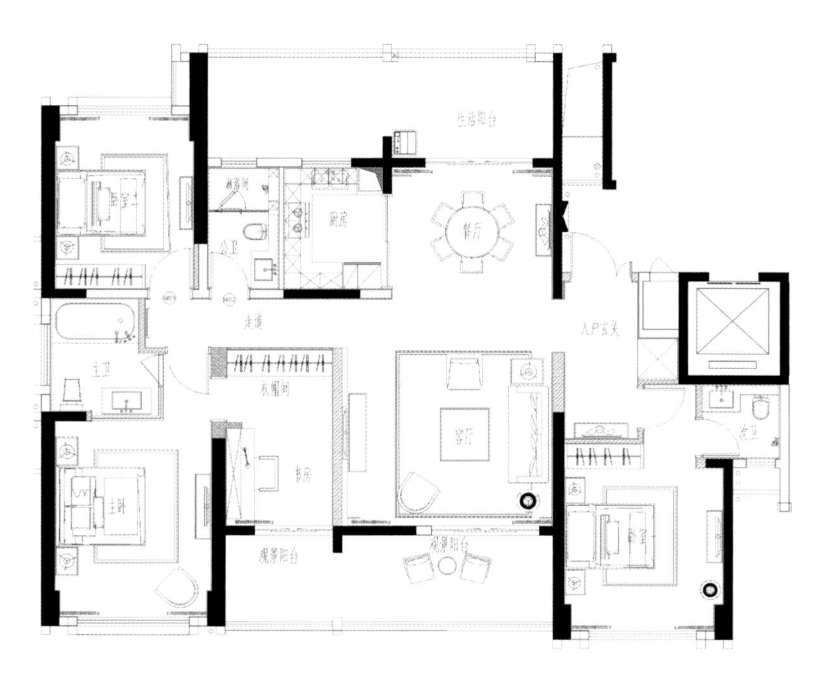

本案的业主所追求的是一种高品质的生活。设计师以后现代的手法打造住宅空间，从视觉呈现到生活设计都进行了独一无二的构想，让空间呈现出具有高端气质同时又独具特色。

精美的材料提升了整体空间的艺术气质，把艺术风格融入现代设计细节中。客厅的设计温馨又大气，色泽莹润的背景墙设计更是为室内带来一缕时尚风情。布艺面料的沙发造型简洁大方，灰色调的颜色使其在空间中不跳跃，沙发上摆放着几个抱枕，舒适感立刻突显出来。在各个空间中，设计师有意识地运用相同的材料形成空间上的连续性。意大利木纹大理石凭借其独特的纹理，吸引着人们的眼球，成为空间中提升质感的重要材质。

设计师在空间中并未采用过于繁复的装饰纹样，而是有意识地简化元素，让设计变得简单、自然，符合业主的心理期望，让典雅的家居空间浸透在一种豪华的设计哲学中。

What the property owner of this project pursues is some high quality life. The designer creates a residential space with post-modern approach, producing unique conceptions towards visual presentations and life design. And that makes the space display some high quality temperament while maintaining its peculiar characteristics.

The exquisite materials uplift the artistic quality of the whole space, with artistic style integrated into the details of modern design. The design of the living room is warm and magnificent. The background wall of bright and smooth luster produces some fashionable feel for the interior space. The sofa of cloth materials is concise and magnificent. The gray color of the sofa makes it be in harmony with the space. There are several bolsters on the sofa which makes it appear comfortable and cozy. For every space, the designer consciously applies some similar materials to make the space consecutive. Based on the peculiar pattern of the Italian wood grain marble, it is very fascinating while uplifting the whole texture of the space.

The designer does not apply much too complicated decorative patterns inside the space. But that he consciously simplifies the elements and makes the design appear simple and natural, in accordance with the psychic expectations of the property owner. This elegant residential space is immerged in some luxurious design philosophy.

LA COURONNE

金地航头样板房

Jindi Hangtou, Show Flat

设计单位：上海乐尚装饰设计工程有限公司
设 计 师：卢尚峰
项目地点：上海
项目面积：253.73 m^2
主要材料：桃花芯饰面板、火焰纹饰面板、新西兰羊毛地毯、皮革、不锈钢

Design Company: Shanghai LESTYLE Decorative Design and Engineering Co., Ltd.
Designer: Lu Shangfeng
Project Location: Shanghai
Project Area: 253.73 m^2
Major Materials: Mahogany, New Zealand Wool Carpet, Leather, Stainless Steel

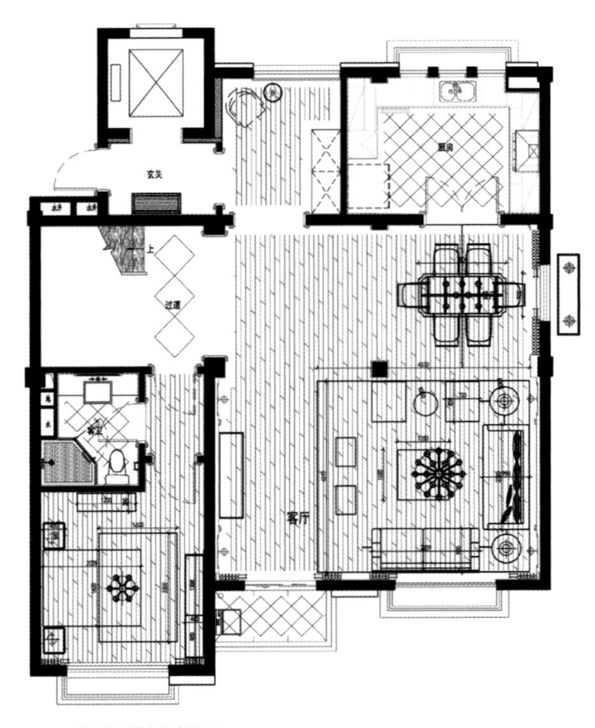

玄关
厨房
过道
客厅

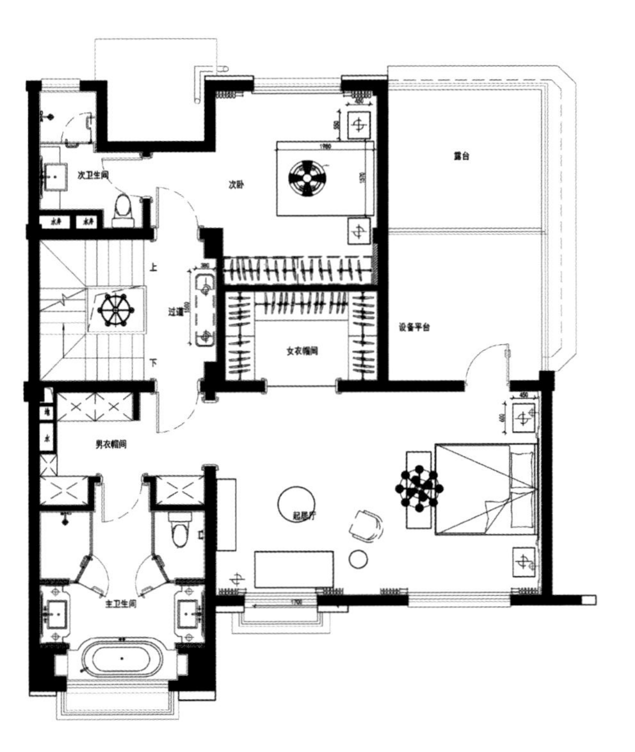

次卫生间
次卧
露台
过道
女衣帽间
设备平台
男衣帽间
起居厅
主卫生间

JACKSON
MICHAEL JACKSON NUMBER ONES
INDOSTR
3W69-22

OLD WORLD INTERIORS

新古典风格强调带给人一种简洁、舒适、自然的心理感受。美式新古典风格是一种新兴的风格，它植根于欧洲文化，同时又摒弃了巴洛克和洛可可风格所追求的新奇和浮华，建立起一种对古典从内而外的认识，强调简洁、明晰的线条和优雅、得体、有度的装饰。像美国人的性格一样，美式新古典风格到处洋溢着追求舒适、欢乐生活的影子。

本案强调空间的独立和联系，以及追求高品质的生活方式的态度。空间的整体配色较为稳重、大气，注重营造唯美、典雅的气息，满足了业主对空间的要求，为业主打造出了与众不同的个性生活空间。设计师将新古典风格的精髓提炼出来，与硬装设计进行了完美的结合。无论是带给人轻松、自然感受的家具，还是精致的配饰，都被和谐地融合在了一起。开放的理念更体现出业主独有的品位和对精致生活的追求。

Neo-Classical Style would always want to give people some concise, comforting and natural psychological enjoyments. American Neo-Classical style is some novel style which is rooted in European culture, while forsaking the novelty and ostentatiousness pursued by Baroque and Rococo style, establishing some understanding towards classicism from interior towards the exterior and emphasizing on concise and clear lines and elegant, decent and proper decorations. Like the characteristics of American people, American Neo-Classical style would set off the comfortable and happy life.

This project stresses on the independence and linking among spaces and the attitude with the pursuit of high quality life. The whole color collocation of the space is comparatively profound and magnificent, emphasizing on aesthetic and elegant atmosphere, meeting with the property owner's requirements for life and creating some peculiar life space for the property owner. The designer extracts the essence of Neo-Classical style, combining in harmony with the hard decorative design. Be it comforting and natural furniture, or delicate ornaments, all are harmoniously integrated together. The open concepts can better represent the property owner's exclusive life taste and pursuit for quality life.

成都南城都汇样板房

Chengdu Nancheng Duhui Show Flat

设计单位：深圳市尚邦装饰设计工程有限公司
设 计 师：潘旭强、刘均如
项目地点：四川省成都市
项目面积：200 m^2
摄 影 师：林力

Design Company:Shenzhen Shangbang Decorative Design and Engineering Co., Ltd.
Designers: Pan Xuqiang, Liu Junru
Project Location: Chengdu in Sichuan Province
Project Area: 200 m^2
Photographer: Lin Li

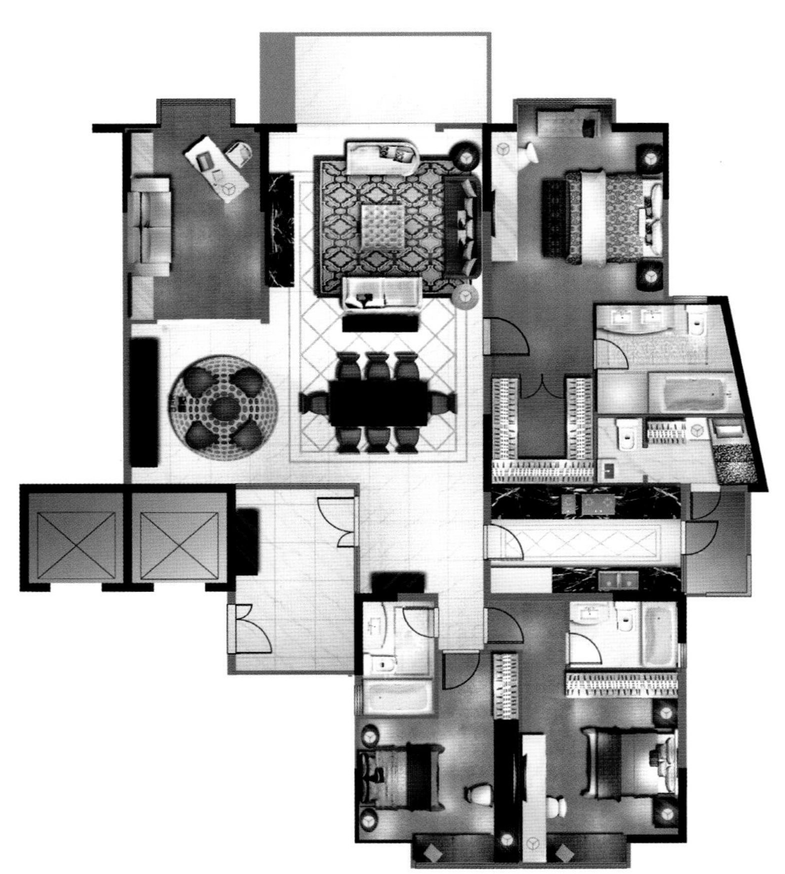

GLAMOUR & LUXURY
GLAMOUR & LUXURY

GLAMOUR & LUXURY
GLAMOUR & LUXURY

本案融入了巴洛克风格的精髓，设计师采用繁复的花纹壁纸来装饰墙面，同时选择巴洛克风格的家具来彰显空间的华丽高贵。室内的设计充分结合现代风格的线性感与时代感来体现空间的气势与协调性，突出夸张、浪漫、激情和非理性、幻觉、幻想的特点。公共空间的布局形式宛如画廊，餐厅墙面的挂画呼应着餐桌的轴线，成为视觉的中心，落地窗边镜面中央镶嵌着的巨大油画仿佛幻境一般，带给人无限的遐想。

设计师敢于打破均衡，利用平面多变的形式强调设计的层次和深度。在本案中，设计师使用各色大理石、宝石、青铜、金等材质来装饰室内空间，彰显华丽、壮观的视觉效果，设计师打破了文艺复兴古典主义的一些模式和原则，运用现代感十足的直线条，让空间的效果更符合现代人的审美观念。整个空间在布局上开敞、连贯，在色彩上明亮、优雅，细节处的装饰物也选用得恰到好处，展现出丰富的视觉效果和浓厚的艺术气息。

This project integrates the essence of Baroque style. The designer applies complicated pattern wallpaper to decorate the wall, while making use of Baroque style furniture to display the magnificence of the space. The interior design sufficiently combines lines sexual feel of modern style and modern feel to display the momentum and harmonious feeling of the space, highlighting exaggerated, romantic, passionate, irrational and visionary characteristics. The layout of the public space is just like gallery. The painting hung on the dining hall's wall corresponds with the axis of the dining table, becoming the visual focus. There is a huge oil painting inlayed in the center of the mirror beside the French window, which is like a fairyland and produces limitless imaginations for people.

The designer dares to break the balance and emphasizes on the layers and depth with varying forms. For this project, the designer makes use of materials such as marble, hade, bronze and gold, etc. to decorate the interior space to demonstrate splendid and magnificent visual effects. The designer breaks through some modes and principles of renaissance classicism and makes use of straight lines of strong modern feel to make the space effects better meet with the aesthetic ideas of modern people. The whole space is broad and consistent in layout, bright and elegant in colors and quite appropriate in the selection of decorative ornaments, displaying rich visual effects and intensive artistic atmosphere.

远雄新都样品屋

Yuanxiong•Xindu Show Flat

设计单位：玄武设计
设 计 师：黄书恒、欧阳毅、许宜真、蔡明宪、胡春惠、胡春梅
项目地点：台湾省台北市内湖区行善路
项目面积：264 m^2
主要材料：银狐大理石、白色冷烤漆、特殊壁纸、南非黑石材
摄 影 师：王基守

Design Company: Sherwood Design
Designers: Huang Shuheng, Ouyang Yi, Xu Yizhen, Cai Mingxian, Hu Chunhui, Hu Chunmei
Project Location: Taibei in Taiwan of China
Project Area: 264 m^2
Major Materials: Silver Fox Marble, White Paint of Baking Finish, Special Wallpaper, South African Black Stone
Photographer: Wang Jishou

美感的迸发，有时来自二维的并置与思辩。无论是东方 VS 西方、现代 VS 古典、科学 VS 玄学，当选定主题风格后，设计师可以运用对立或结合两极的元素，巧妙拮取装饰语汇，激荡出新奇的诗意与美学。

巴洛克风格是表现力量与富足的装饰风格。本案的现代巴洛克风格则取其精神，而去其繁复装饰，在华丽变幻的风格中，用色不再夸张，描金只在细节中含蓄表露。设计师以新古典风格为主轴，将繁复的古典语汇简化，并融入英式的庄重气息，进而在这户庄园般的宅邸中，糅合出简约雅致的新古典风格传奇。

入口的大理石双圆图案，对应玄关的镂空圆与线相嵌屏风，大气且低调。古典的语汇以圆弧的温润成为浸透视觉的享受。圆形雕饰的顶棚下，圆桌、收边角的椅，深褐色烛台缀链吊灯，映照在餐厅背景墙的墨镜上，华丽却不奢华。

巴洛克时代的壁炉到现代被电视机所取代，电视反映着现代生活的中心，黑白岗石强调的电视柜、长沙发、更衣室门片，简单就是所有真正优雅的基调。缓步在此美丽庄园中流连，柔和而经典的气息，似乎在空间中轻盈跳跃。如同黑键与白键的交错，

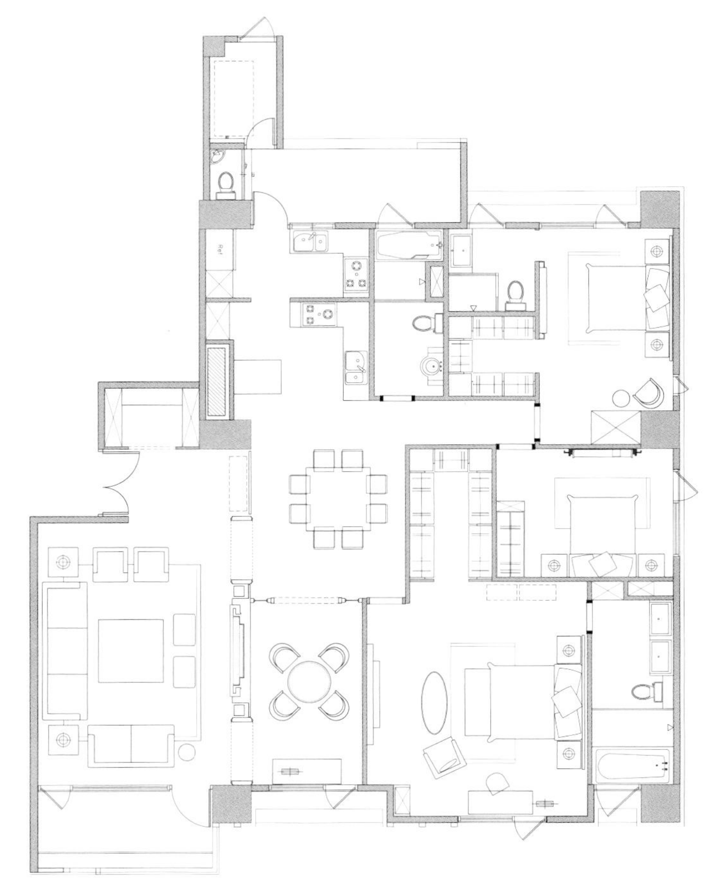

优雅跳跃的银边则如休止符，透过如同音符般的意象呈现，此起彼落，在空间中弹奏一曲轻柔隽永的香颂名曲，不只让主人留恋依傍，更让来访的每一位客人，流连忘返，聆赏醉心。

本案设计师特别在公共空间中，以黑与白的低彩度设计带出古典的气息。以客厅为例，白色布沙发与黑色镶白边的主人椅，在空间中进行着优雅的对话。除此之外，设计师为了不让空间过于单调，将两只进口马毛单椅陈设其中，既带出质感，又有画龙点睛之效。

床尾的壁炉造型掩饰着电视的摆放处，缎面光泽的银底黑花主人椅，点亮这一方低彩度的空间。值得一提的是主卧的更衣室，设计师刻意打造有如高档精品店般的展示空间，让业主在试衣时，也能尽情在这一方私密空间中旋转、舞动。床头壁板为简化的英式古典造型，有巴洛克风格的精细对称，却没有其雕琢的繁复沉重。梳妆区也维持一贯的黑白对比带来的惊艳。家饰皆细心描绘镶边，创造低调的精致风格。主卧卫浴为干湿分离，有浴缸、花洒、双洗脸盆等，功能一应俱全。银色雕花的镜面边框，更成为这一方盥洗空间中的焦点。次卧室有卫浴与更衣室，风格色调也维持一贯的娴静优雅。

The explosion of aesthetic beauty sometimes comes from the two dimensional juxtaposition and speculation. Be it oriental VS west, modern VS classical, or scientific VS metaphysical, when a theme style is decided, the designer can make use of contrasting or integrating elements, ingeniously selecting decorative language and agitating novel poetic and aesthetic sensations.

Baroque style is some decorative style displaying strength and quality. This project takes the essence of modern Baroque style and removes the complicated decorations. In the magnificent and varying style, the colors are not that exaggerated. The designer takes neo-classical style as the main axis, simplifies complicated classical language and integrates inside English solemn atmosphere. Thus inside this manor like mansion, the legend of neo-classical style is created.

The marble graphics at the entrance are magnificent and understated, echoing with the hallway screens. With its circular arc warmth, classical languages become visual enjoyment. Under the ceiling with round decorations, round tables, chairs and dark brown drop lights are reflected in the mirror on the background wall of the dining hall, magnificent but not luxurious.

Fireplace of Baroque time is replaced by modern TV which reflects the focus of modern life. TV stand, long sofa and

MODERN GLAMOUR
MODERN GLAMOUR

changing room are emphasized with black and white marble stone. Simplicity is the real elegant tone. When you are walking slowly in this nice manor, some soft and classical atmosphere seems to dance in light and graceful way inside the space. It is like the intersection of black keys and white keys. Graceful and bouncing silver lining is like some pausing note, displayed with some musical note phenomenon. It is like some gentle and meaningful chanson musical composition inside the space. This space does not only make property owner reluctant to leave, but also make every incoming visitor enjoy himself so much as to forget to leave.

The designer of this project tries to create low saturation of black and white with classical atmosphere. To take the living room as an example, sofa of white cloth and black chair of white lining perform graceful dialogue in the space. Other than that, in order that the space does not appear much too monotonous, the designer placed two imported horsehair chairs inside, creating texture while making the finishing point.

Upon entering the master bedroom, you can find the interlacing variation of black and white, creating cozy and restrained intricate style. The TV stand is hidden behind the furnace sculpture at the end of the bed, with black chair of satin luster highlighting this space of low saturation. What deserves mention is the changing room inside the master bedroom. The designer deliberately creates some presentation space which is like some high-standard boutique store, thus the designer can dance and spin at heart's contents while changing clothes. The wallboard at the head of the bed has simplified English style classical modeling, with exquisite symmetry of Baroque style, but no intricate complexity and heaviness. The dresser also maintains the black and white contrast as usual. The home decorations' lining are all carefully decorated, creating low-key intricate style. The master washroom separates dry area from the wet area, with bathtub, sprinkler, two basins, etc., performing complete functions. Mirror frame of silver carving becomes the focus of the washing space. The secondary bedroom has washroom and changing room, with style and color tones maintaining the consistent tranquility and elegance.

名城国际——关于欧式岁月的构想

Mingcheng International – Conception about European Years

设计单位：福建品川装饰设计工程有限公司
设 计 师：林新闻、林锦
项目面积：180 m^2
主要材料：鳄鱼皮、皮革、绒布、缎布、木饰面板、仿古大理石、罗马洞石、地毯
摄 影 师：周跃东

Design Company: Fujian Pinchuan Decorative Design and Engineering Co., Ltd.
Designers: Lin Xinwen, Lin Jin
Project Area: 180 m^2
Major Materials: Crocodile Leather, Leather, Flannel, Satin, Wood Veneer, Antique Marble, Roman Travertine, Carpet
Photographer: Zhou Yuedong

通常来说，充足的室内面积可以让古典风格的家居尽情展示其大气、华丽、高贵的气质，这也是欧式风格之所以受成功人士青睐的重要原因。在这套样板房中，设计师以欧式新古典凸显空间的堂皇与奢华之美，同时也融合了多元化的简约元素，展现出居室低调的华丽。沉稳的色调、通透的空间、精致的细节与线条造型，让这套样板房不只拥有华美，也拥有温婉、多情的一面。

客厅装修无疑是设计师花费时间和精力最多的空间，也是整个居家格调的重墨之笔。设计师以大面积的落地玻璃窗为起始，客厅、餐厅之间没有复杂的隔断，空间框架继承了实木材料的古典美，玻璃、金属等现代材质也被运用其中，一个个木质框架将四周墙壁分割成若干几何板块，塑造了经典的西式设计。设计师还秉承了新古典主义一如既往的舒适，空间整体以暖色调为主，搭配冷色系镶雕花金属的布艺沙发和茶几，增添几分沉静，充满质感的混搭和丰富的层次中所包容的大度使居室品质更具大家风范。设计师通过对墙体的刻画凸显空间的层次感，一面背景墙，无论轻描淡写或浓妆艳抹，无论是雕花纹理还是书香挂画，都能通过华丽的装饰与浓烈的色彩烘托出整个空间的氛围。

设计师精心布置了四间极具浪漫特色风情的卧室。在填充元素及主体基调不变的基础上，通过色彩、材质、造型、布局改变，以及红、棕、白、银四个主题色调的风格微调，展现了百变的华丽。床品皆选用暖色系，搭配各种灯饰、边几、窗帘、抱枕，营造出一室的暖意和温情，显得格外的古典和奢华。值得一提的是，每一个房间都设置了飘窗或落地窗，抬眸一望，室外的浩瀚江景一览无余，将窗外景色引入室内，简单却又不失华贵气息。

Generally speaking, resplendent interior space can make the furniture of classical style display its magnificent, gorgeous and noble temperament. This is also an important reason for European style to enjoy the favor of successful people. For this model house, the designer uses European neo-classical style to highlight the grand and luxurious beauty of space. While at the same time, the design integrates diversified concise elements to display the low-key magnificence of the residential space. Sedate color tones, transparent space and exquisite details and line formats make this sample house not only possess splendor, but also display the warm and passionate side.

Living room is definitely the space where the designer spent the most time and energy. It is also the focus of the style of the residence. The designer starts from large area of French glazed window. There are no complicated partition between the living room and the dining hall. The space structure inherits the classical beauty of solid wood materials. Modern elements such as glass and metal are also included inside. Some wooden frames separate the surrounding wall into geometric plates, establishing classical western design. The designer also inherits the comfort of neo-classical style and centers the whole space on warm color tones. Accompanied with cloth sofa and tea table of cold color metal carving, the space is added with some tranquility. Mix and match of texture and generosity in the rich layers make the whole residential space with mater style. Through inscription of designers, the designer highlights the layers of the space. This background wall can set off the atmosphere of the whole space with magnificent decorations and intensive colors, be it understated or magnificent, be it carving pattern or paintings.

The designer elaborately set four bedrooms of romantic characteristics and charms. On the basis of instilling elements and subject fundamental keys, through changes of colors, materials, formats and layout and minor adjustment of four theme color tones, red, brown, white and silver, the designer displays the magnificence varying all the time. The bedding accessories all apply warm colors, accompanied with lights, lining, curtain, sofa pillows, etc., hence creating a room of warmth and presenting peculiar classical and luxurious style. It is necessary to speak of the fact that there are windows or French windows inside every room. When you look outside, you can possess a broad view of vast river sceneries. Thus the exterior sceneries are introduced inside, simple but noble at the same time.

金色年华

Golden Age

设计单位：易成设计　　Design Company: Yicheng Design
设 计 师：郑悦音　　Designer: Zheng Yueyin
摄 影 师：李玲玉　　Photographer: Li Lingyu
撰　　文：林飞飞　　Composer: Lin Feifei

金色，是一种最辉煌的颜色，更是大自然中至高无上的颜色。它是太阳的颜色，代表着温柔与幸福。同时，也拥有照耀人间、光芒四射的魅力。

金色能够给人极醒目的视觉感受。它具有一个协调周边颜色的特性——在各种颜色的配置不协调的情况下，使用了金色就会使它们立刻和谐起来，并产生光亮感以及华丽、辉煌的视觉效果。

在本案中，金属的质感随处可见，一进大厅，立刻就能感受到一股奢华之风。金属镶边的沙发组合与茶几的颜色相统一，同时又与背景墙上巨大的装饰品颜色相协调，使整个空间浑然一体、张力十足。顶棚的吊灯采用欧式风格，工艺繁复、精致，色彩张扬、大胆。客厅旁边用隔断打造出一个就餐区，狭长的餐桌与复古的椅子沿袭了整体的简欧风格。在灯光的渲染下，空间中溢满金色的光芒，华丽精致。

Gold is some glorious color, which is also the supreme color in the grand nature. It is the color of sun which represents mildness and happiness. While at the same time, it has the charms of lighting up the world with its rays of light.

Gold can leave people with striking visual impressions. It has the qualification of harmonizing the surrounding colors. Under the circumstances of all colors not in harmony with each other, gold can make them coexist in harmony all at once, while producing light, magnificent and splendid visual effects.

For this project, the texture of metal is everywhere. Upon entering the living room, you can perceive this wind of luxury. The sofa of metal lining is in harmony with the color of the tea table, while coordinating with the grand decorative objects on the background wall, making the whole space an integrity and with intensive tension. The ceiling chandeliers apply European style, with complicated and exquisite techniques and bold and audacious colors. There is a dining area beside the living room created with a partition. The long dining table and the archaic chairs inherit the concise European style. Set off by the lights, the space is full of gold rays and magnificence.

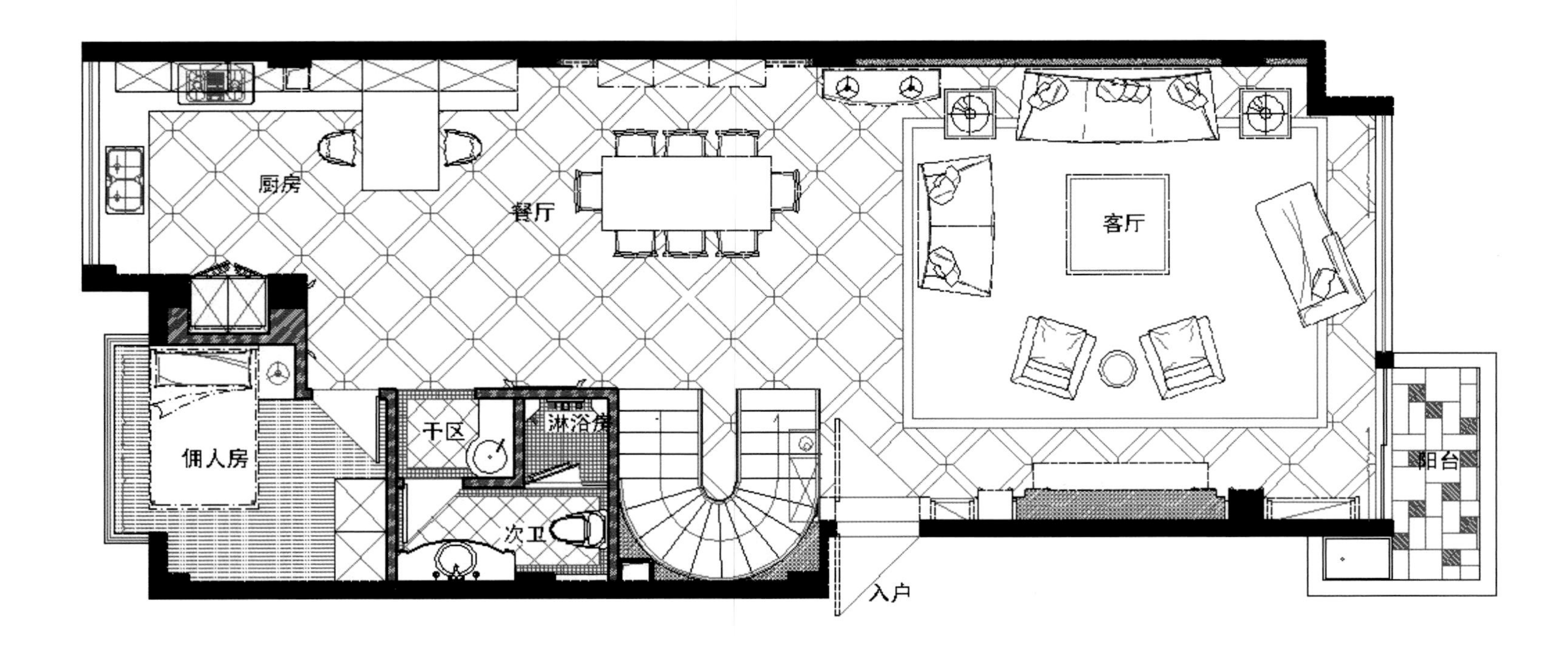
厨房
餐厅
客厅
干区
淋浴房
佣人房
次卫
阳台
入户

中联·天御样板房A

Zhonglian•Tianyu Show Flat A

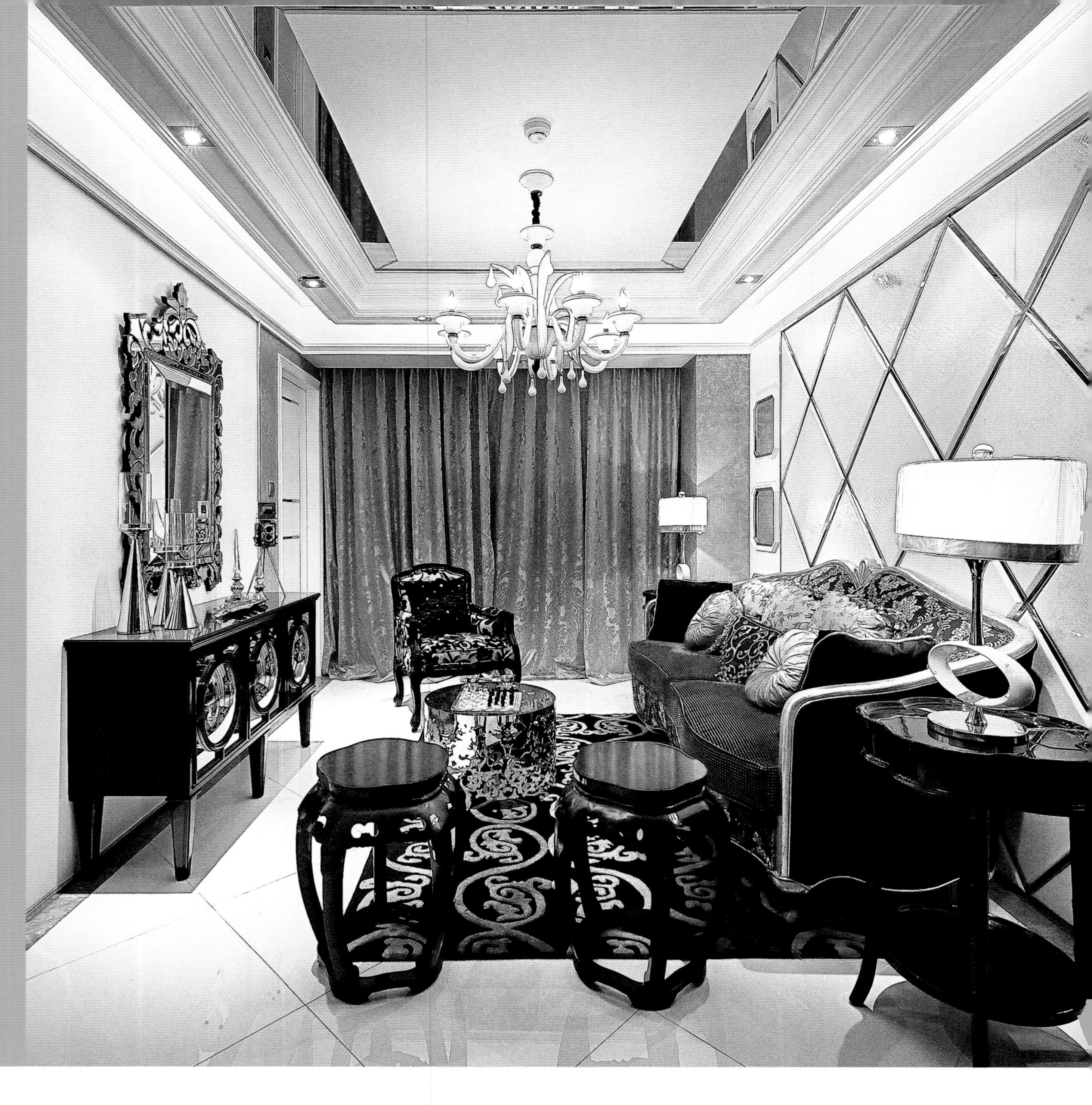

设计单位：福建国广一叶建筑装饰设计工程有限公司
设 计 师：何华武、郭礼桑
项目地点：福建省福清市
项目面积：50 m^2
主要材料：饰面板、壁纸、皮革硬包、艺术陶瓷锦砖、镜面、木地板、地毯、镜面不锈钢
摄 影 师：李玲玉

Design Company: Fujian Guoguangyiye Construction Decorative Design and Engineering Co., Ltd.
Designers: He Huawu, Guo Lishen
Project Location: Fuqing in Fujian Province
Project Area: 50 m^2
Major Materials: Panel, Wallpaper, Leather Hard Roll, Artistic Ceramic Mosaic Tile, Mirror Surface, Wood Flooring, Carpet, Mirror Surface Stainless Steel
Photographer: Li Lingyu

本案的欧式新古典风格是意大利风格的一种延伸，与传统的古典风格相比，它不再呈现无比庞大、厚重的气质，而是向轻盈、小巧发展，充满时尚感。如果说古典欧式风格多采用大理石装饰，以金色色调为主，那么，新古典风格则更多采用银箔、白色材质，搭配不锈钢和玻璃的质感，充满小资情调，高贵而不浮华。

欧式新古典的家具一反纯粹古典主义的厚重、沉闷，体量不再一味求大求重，造型也不再繁复；材质不再以木制、皮质为主，而是多种材料并行，尤以银箔亮漆使用较多，主要框架多采用不锈钢材料，营造出一个轻灵、透亮的空间；布艺面料多以丝绒为主，既显得华贵，又彰显浪漫的品质。新古典风格有一个最为明显的标志，即壁纸、窗帘风格一致，多采用小勾花为主的矢量图花纹。不同于乡村风格洒满墙壁的碎花，带来的是犹如大自然一般的清新和浪漫，新古典风格的花纹造型简单，排列整齐，不断重复，规则却不死板，既有现代感又显得富丽堂皇。设计在充分顾及功能的同时，还必须体现美感，要具备鲜明的风格特征。传统古典风格的灯饰，以金色为主，造型繁复，层层叠叠，注重细枝末节，着重展现贵族的豪奢。新古典风格的灯具造型简洁，但依然保留了古典韵味，不但采用金属材质，还饰以玻璃、镜面等，平添了一丝时尚。如果说古典风格的台灯灯罩以梯形为主，那么直筒圆柱则是新古典风格台灯的醒目特色。地毯搭配羊毛质感，体现出柔美的感觉。瓶子、烛台、镜框都是常见的装饰品，和窗帘、壁纸一样，尽可能带有小勾花的花纹造型。总体来说，新古典风格属于一种大众风格，无论是偏古典还是偏现代，都可以达到简约精致的效果，既能显得时尚，也能做到温馨。

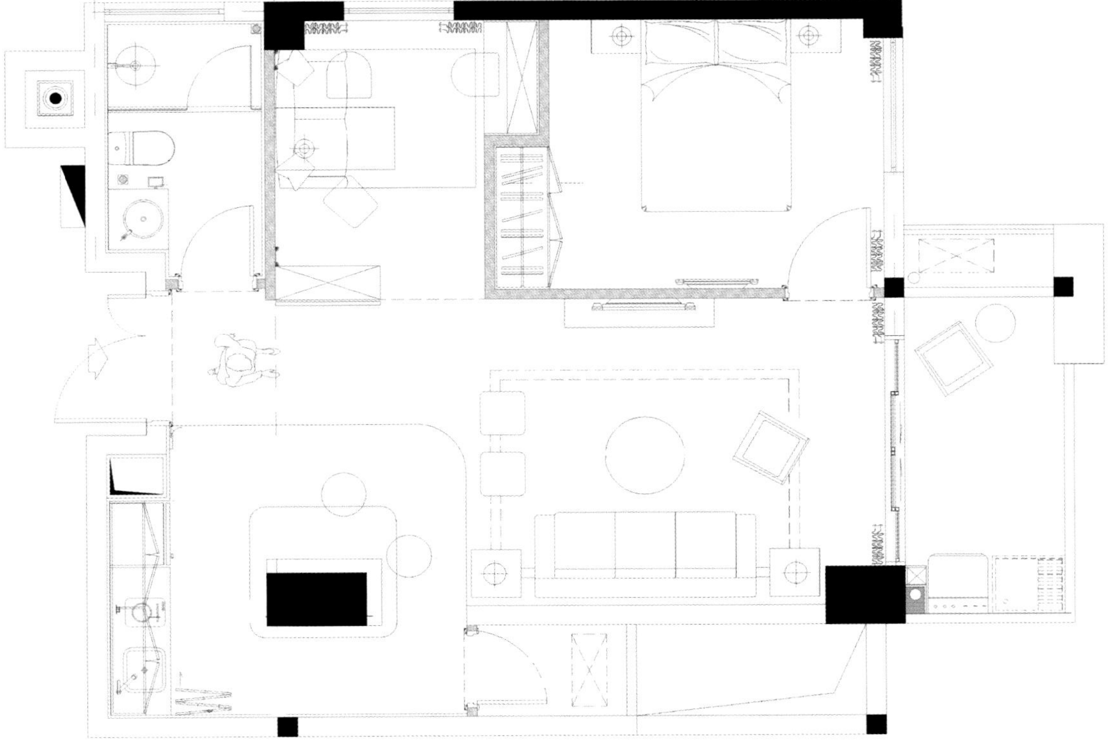

The neo-classical style of this project is the extension of Italian style. Compared with traditional classical style, it does not try to display bulky or heavy temperament, but develops towards light, tiny and fashionable style. If classical style mainly applies marble decorations with gold color tone, neo-classical style more applies silver foil and white materials, accompanied with stainless steel and glass, full of petty bourgeoisie charms, noble but not showy at all.

European neo-classical furniture goes against the heaviness and dreariness of pure classical style, with no grand and heavy scale, nor complicated format. Materials no longer focus on wood and leather only, but with many materials, especially silver foil and light paint. The main frame mainly applies stainless steel, thus creating a lively and transparent space. Fabric materials center on velvet, displaying magnificence and romantic quality. The most obvious quality for neo-classical style is the consistent style of wallpaper and curtain, mainly making use of vector diagram pattern. Scattered flower pattern on the wall different from countryside style creates some freshness and romance just like the grand nature. Neo-classical style flower pattern has simple formation, tidy in layout, but with no repetitions. It appears orderly and active, with modern feel and magnificent atmosphere. While taking functions into account, the design must display aesthetic feel, with brisk style characteristics. Lighting accessories of traditional classical style center on gold color, with complex format of layers upon layers. The design emphasizes on details and sets to display the luxury of aristocrats. Neo-classical style lighting accessories are concise, but maintain classical charms. It applies metal materials with glass and mirror surface decorations, with some fashionable sensation. If classical lampshades are mainly trapezoid, cylinder style is the main characteristic for neo-classical style lamps. The combination of carpet and wool displays some soft sensations. Vase, candlestick and mirror frame are all common decorative objects. They all possess little flower pattern, just like curtain, wallpaper, etc. Generally speaking, neo-classical style is some public style. Be it classical or modern style, all can achieve concise and exquisite effects, being fashionable and warm at the same time.

中联·天御样板房B

Zhonglian•Tianyu Show Flat B

设计单位：福建国广一叶建筑装饰设计工程有限公司
设 计 师：何华武、段玮
项目地点：福建省福清市
项目面积：50 m²
主要材料：饰面板、壁纸、镜面、木地板、地毯、镜面不锈钢
摄 影 师：李玲玉

Design Company: Fujian Guoguangyiye Construction Decorative Design and Engineering Co., Ltd.
Designers: He Huawu, Duan Wei
Project Location: Fuqing in Fujian Province
Project Area: 50 m²
Major Materials: Panel, Wallpaper, Mirror Surface, Wood Floor, Carpet, Mirror Surface Stainless Steel
Photographer: Li Lingyu

SACON

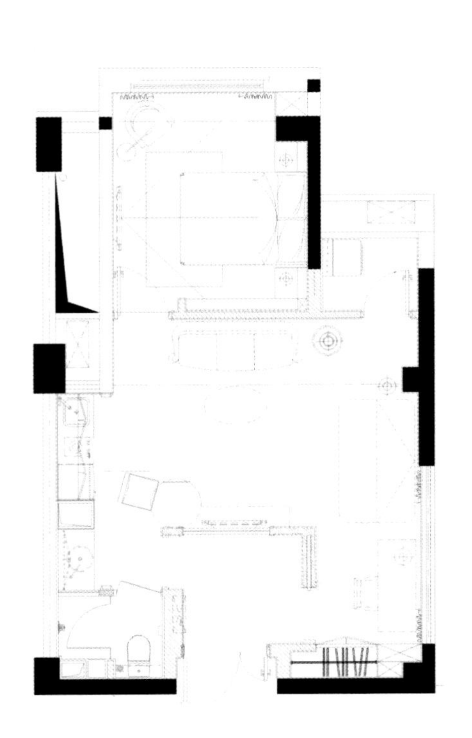

欧式新古典风格的设计最适合营造典雅、高贵的气质和浪漫的情调。本案借着室内空间的解构和重组，缔造出一个令人心驰神往的写意空间。在家居装饰中，不同的风格可以演绎出各种各样的风情，蕴涵着千姿百态的家庭生活乐趣。任何装修风格的前提都是保障生活的舒适性，空间要体现出业主丰富的生活阅历，在空间布局整体色彩运用上有丰富的理念。整个生活空间彰显出一种安逸的舒适性，又展现出流行设计元素的搭配，巧妙的软装设计让人耳目一新。

优雅的气质和生活品质感是设计表现的主旨。欧式家居总是给人一种沉稳大气的感觉，在欧式新古典风格的家中，犹如置身于中世纪的古堡之中。生活在此，感受到的更多是惬意和浪漫，整体空间具有强烈的西方审美气息，设计师用华丽的装饰、浓烈的色彩、精美的造型达到雍容华贵的装饰效果。

European neo-classical style is most appropriate for creating elegant and noble temperament and romantic artistic conceptions. Based on deconstruction and recombination of the interior space, some exciting poetic space is created. For the decoration of home furnishing, different styles can produce various artistic conceptions, with family life interests differing in thousands of ways. The premise for any kind of decoration style is to guarantee the comfort of living, with space which can display the abundant life experiences of the property owner and space layout which has rich philosophies in the whole color applications. The whole residential space displays some easeful comfort, with combination of design elements in fashion. The ingenious soft decoration design makes people find themselves in an entirely new world.

Elegant temperament and life quality are the motif that design tries to express. European home furnish would always leave people with sedate and magnificent impressions. Inside a home of neo-classical European style, it is like being in an old castle of medieval time. Living here, people would get more comfort and romance. The whole space displays some intensive western aesthetic atmosphere. The designer makes use of gorgeous decorations, strong colors and exquisite format to achieve the noble and aristocratic decorative effects.

华地公馆 A1-1 样板房

Huadi Mansion, A1-1 Show Flat

设计单位：深圳市墨客环境艺术设计有限公司
设 计 师：王勤俭
项目地点：安徽省合肥市
项目面积：168 m^2
主要材料：黑金花、橙皮红、金柠檬黄大理石，瓷砖，金箔，米白洞石，球纹桃花芯木饰面板，艺术拼花陶瓷锦砖，红檀香实木地板，壁纸，水晶灯，银镜

Design Company: Shenzhen Moke Environmental Art and Design Co., Ltd.
Designer: Wang Qinjian
Project Location: Hefei in Anhui Province
Project Area: 168 m^2
Major Materials: Black Gold Marble, Orange Red Marble, Gold Lemon Yellow Marble, Ceramic Tile, Gold Foil, Creamy White Travertine, Wood Veneer, Ceramic Mosaic Tile, Sandalwood Flooring, Wallpaper, Crystal Light, Silver Mirror

本案原始格局是一个 168 m² 的三房两厅两卫，通过设计优化把原来的空中花园引入室内，增加了多功能房和健身区。主卧室内将原来的阳台改造成开放式衣帽间，结合玻璃材质使其变得宽敞通透，另外将主卧室的入口改在过道侧面，在过道端头形成入户对景，符合中国人对建筑风水的要求。

本案运用了新古典的设计手法，将古典欧式繁复的装饰凝练得更为简洁精雅，为硬而直的线条配上温婉雅致的软性装饰，将古典美注入简洁实用的现代设计中，并与现代的材质相结合。让古典的美丽穿透岁月，在我们的身边活色生香，呈现出古典而简约的新风貌，体现一种多元化的思考方式；将怀古的浪漫情怀与现代人对生活的需求相结合，兼容华贵典雅与时尚现代，反映出业主个性化的美学观念和文化品位。设计师以白色、米黄、香槟金色和暗红为主色调，少量白色糅合，使色彩看起来明亮、大方，使整个空间给人以开放、宽容的非凡气度，高雅而和谐，让人丝毫不觉局促。

The original layout of this project is a space of 168 m² with three bedrooms, two halls and two bathrooms. The former hanging garden is introduced into the space with optimized design, with added multi-functional rooms and body-building area. The former balcony inside the master bedroom is designed into an open cloakroom and appears capacious and transparent with glass materials. Other than that, the entrance of the master bedroom is set at the side of the corridor, forming some opposite scenery at the end of the corridor, meeting Chinese people's requirements for architectural style.

This project applies neo-classical design approach and makes classical European complicated decorations concise and elegant, assigns warm and elegant soft decorations for

hard and straight lines and applies classical beauty into practical modern design while combined with modern materials. Classical beauty seems to penetrate through life and shows its vividness around us, displaying classical and concise new appearance and diversified thinking patterns. Romantic emotions with meditation on the past are combined with modern people's requirements for life, including aristocracy, elegance, fashion and modernity, reflecting individual aesthetic viewpoints and cultural tastes of property owners. The designer takes white, beige, champagne gold and dark red as the tone colors, mixed with some white colors, which makes the colors appear bright and generous. The whole space has some open and uncommon temperament, which is elegant and harmonious and does not produce any cramped impressions.

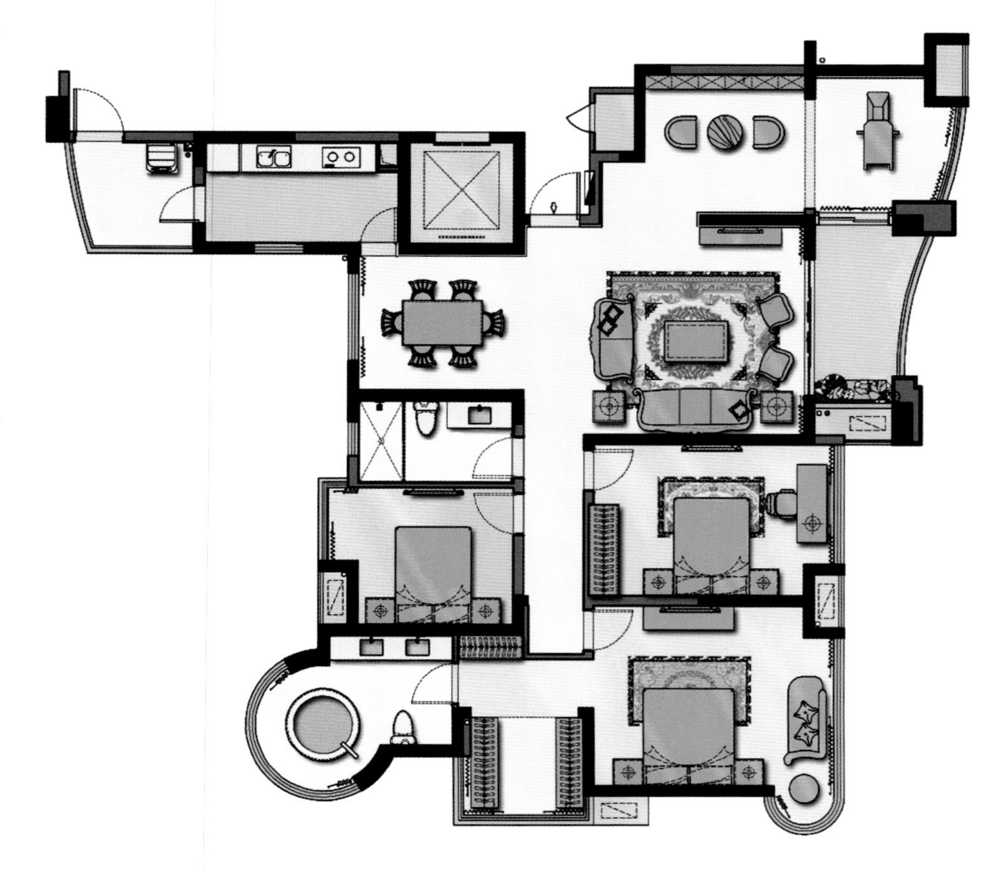

华地公馆B1-1样板房

Huadi Mansion, B1-1 Show Flat

设计单位：深圳市墨客环境艺术设计有限公司
设 计 师：王浩
项目面积：127 m^2
主要材料：白玫瑰、香槟红、黑金花、雅士白大理石，不锈钢，回纹陶瓷锦砖，车边银镜，水银镜磨花，壁纸，绒布软包，密度板锣花，哑银箔纸

Design Company: Shenzhen Moke Environmental Art and Design Co., Ltd.
Designer: Wang Hao
Project Area: 127 m^2
Major Materials: White Rose Marble, Champagne Red Marble, Black Gold Marble, Jazz White Marble, Stainless Steel Ceramic Mosaic Tile, Decoration Mirror, Wallpaper, Velvet Soft Roll, Dull Silver Paper

本案想传递一种低调奢华的生活方式，将欧式生活的贵气与精致带入生活，却又无金碧辉煌的浮华，无论坐在客厅还是花园小憩，一杯浓浓的咖啡，一本小说，都是一种意境，一种浪漫……

对格局合理的改造使空间整齐有序、功能完整。经过调整后，开门处正对入口案台，左侧为客厅，右侧为钢琴厅（把餐厅调到挑空花园的位置）。布局双厅的概念，再结合磨花玻璃、磨花镜的虚拟反射，使空间奢华大气。利用另一挑空花园设置一间客房，让大宅式住宅的主客私密性得到了保证。

整套大宅的设计是奢华主义与现代浪漫主义的完美结合，采用流线形与直线相互搭配，更显家具的现代奢华、精美华丽。明亮的色彩在室内相互交错，豪华的家饰与个性的壁纸相互呼应，此情此景充斥着浪漫的诱惑力与华贵感，从而表现出大气的王者风范。

What this project tries to convey is some low-key and luxurious lifestyle, bringing the aristocracy and delicateness of European life into everyday life, but with no magnificent ostentatious quality. No matter you sit inside the living room or take a rest in the garden, with a cup of strong coffee or a novel book in hand, it is some artistic conception, some romantic enjoyment...

The proper transformation towards the layout makes the space appear tidy and orderly, with complete functions. After readjustment, the entrance faces the counter, with living room on the left and piano hall on the right, while the former dining hall is set in the area with a garden. The space is made luxurious and grand with the concept of double living rooms, combined with virtual reflection of polished glass and convex mirror. There is a guest room in the other garden area, guaranteeing the master guest room's privacy of the mansion residence.

The design of the whole residence is the perfect combination of luxury fashion and modern romanticism. With the mutual combination of streamline forms and straight lines, the furniture

IMPERIAL CROWN

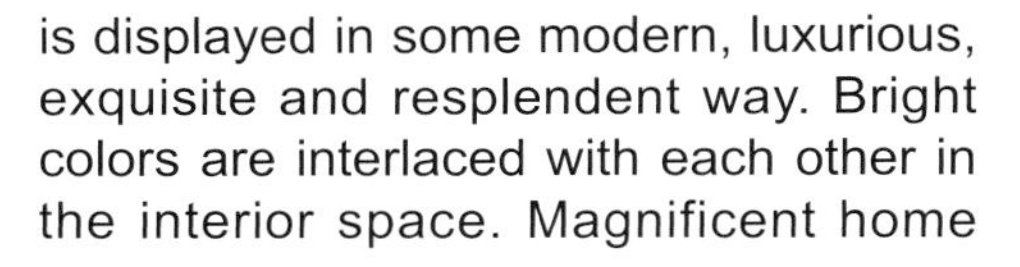

is displayed in some modern, luxurious, exquisite and resplendent way. Bright colors are interlaced with each other in the interior space. Magnificent home decorations and individual wallpaper correspond with each other. There is some romantic attractiveness and gorgeous feelings in this scenery and this situation, presenting some gorgeous king style.

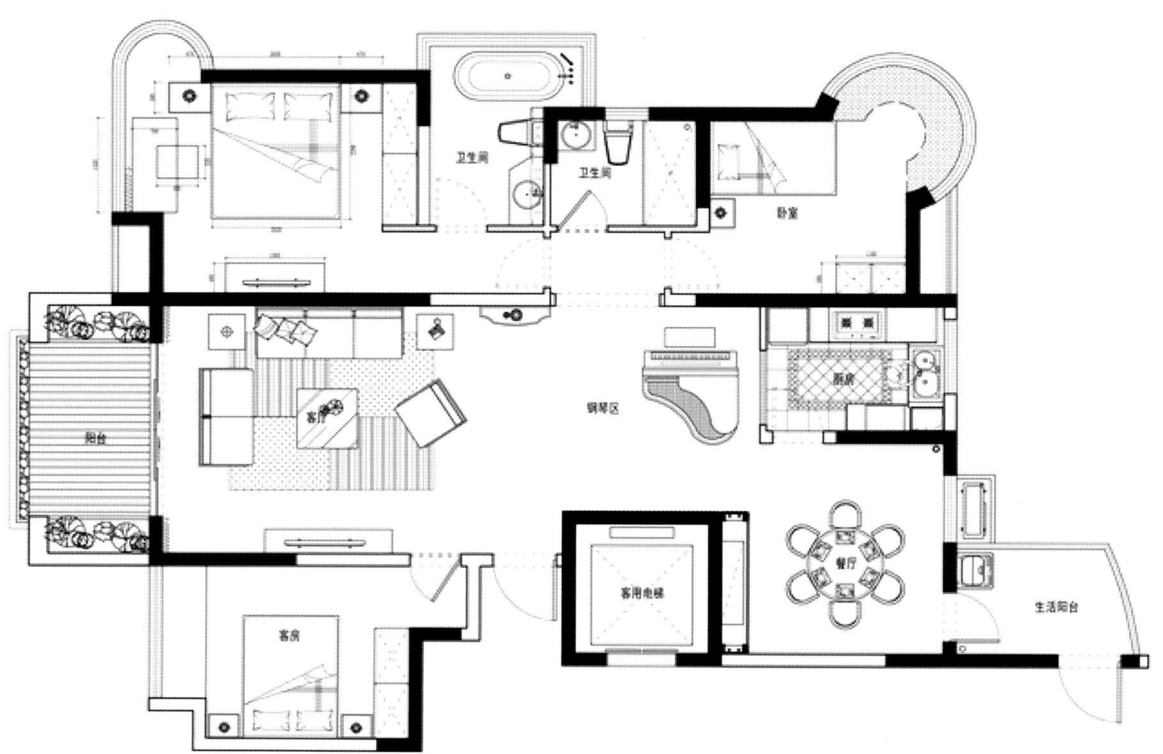
卫生间
卫生间
卧室
客厅
钢琴区
厨房
阳台
餐厅
客用电梯
生活阳台
客房

东方银座中心城样板房

Oriental Plaza Central Town, Show Flat

设计单位：深圳市盘石室内设计有限公司
设 计 师：吴文粒
项目地点：辽宁省沈阳市
项目面积：160 m^2
主要材料：维也纳米黄大理石、壁布、香槟金装饰线条

Design Company: Shenzhen Panshi Interior Design Co., Ltd.
Designer: Wu Wenli
Project Location: Shenyang in Liaoning Province
Project Area: 160 m^2
Major Materials: Vienna Beige Marble, Wallpaper, Champagne Gold Decorative Lining

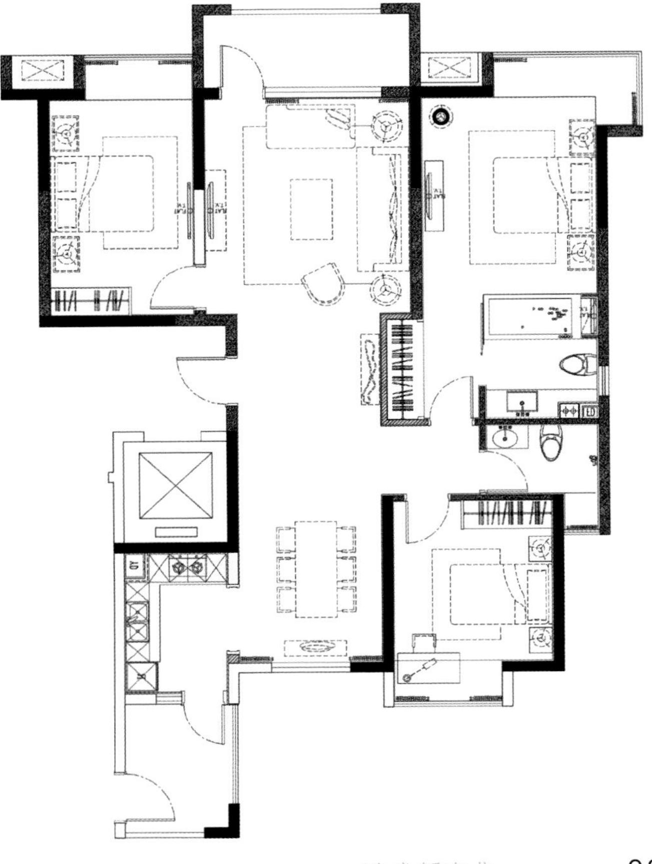

通过传统的文化特征、现代的居住环境体现出对“贵族精神”的崇尚，然而随着时尚观念的不断发展，这种昔日以奢华取胜的“贵族精神”逐渐偏向于简约的潮流，在保留原有生活品位的同时兼具美观性和实用性。

设计师从生活的实际角度出发，在空间的设计中，设计师保留着对欧式古典风格的欣赏，同时简化了古典风格过于华丽却不实用的装饰，让设计在提升生活层次的同时也变得实用起来。棕色系的色调让室内空间充满沉稳大气之感，香槟金装饰线条为空间添加了几抹亮色，在璀璨的水晶吊灯的照耀下，室内空间显得和谐、典雅。简约的墙面和地面处理，结合传统、典雅的欧式家具，在设计师精心挑选的软装饰物的搭配之下，使整个空间的气质完美展现。

The advocation for noble spirits are displayed in the traditional cultural features and modern residential environment. Together with the development of fashion and attitude, this spirit with past luxury tends for concise trends. While maintaining the original life taste, the design displays aesthetic quality and practical characteristics.

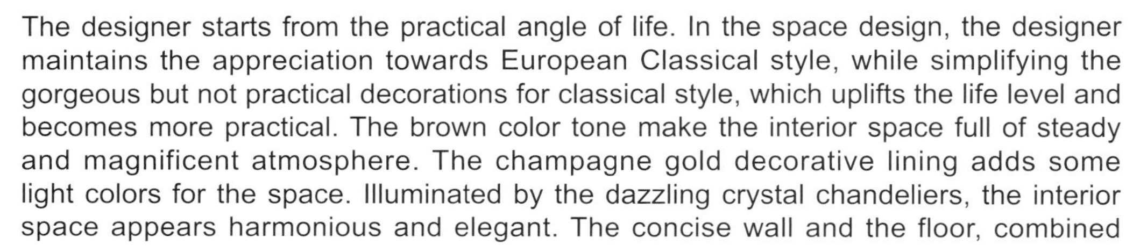

The designer starts from the practical angle of life. In the space design, the designer maintains the appreciation towards European Classical style, while simplifying the gorgeous but not practical decorations for classical style, which uplifts the life level and becomes more practical. The brown color tone make the interior space full of steady and magnificent atmosphere. The champagne gold decorative lining adds some light colors for the space. Illuminated by the dazzling crystal chandeliers, the interior space appears harmonious and elegant. The concise wall and the floor, combined with traditional and elegant European furniture, the temperament of the whole space is displayed in some perfect way with the soft decorative objects carefully selected by the designer.

远中风华七号楼

Yuanzhong Fenghua, Building No. 7

设计单位：玄武设计
设 计 师：黄书恒、欧阳毅、陈怡君
软装设计：玄武设计
项目地点：上海市静安区
项目面积：267.03 m^2
主要材料：黑金花、黑云石、金洞石、浅金锋、卡拉拉白等大理石，黑檀木，银箔，金箔，壁纸，明镜，墨镜，镀钛板
摄 影 师：王基守

Design Company: Sherwood Design Group
Designers: Huang Shuheng, Ouyang Yi, Chen Yijun
Soft Decoration Design: Sherwood Design
Project Location: Jing'an District in Shanghai
Project Area: 267.03 m^2
Major Materials: Black Gold Marble, Black Marble, Gold Travertine, Carrara Marble, Ebony, Silver Foil, Gold Foil, Wallpaper, Mirror, Titanium Plate
Photographer: Wang Jishou

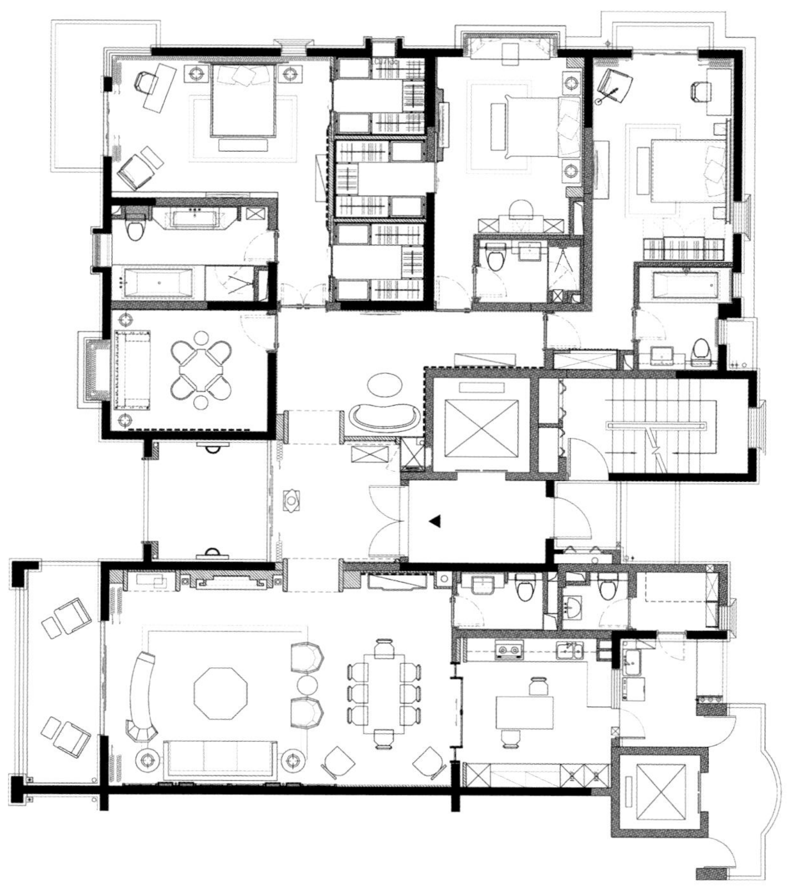

理性缀点　鎏金岁月

设计的未来，
在于把对过去的尊崇，
巧妙无形地蕴藏于对明日的憧憬。

—设计师 Jeffrey Bilhuber

真正的名流豪邸，不在于呈现镶金包银的俗韵，而在于展现大格局、大气度的绝世风华。六角形的浮雕顶棚，大胆布满整个客厅，让西方的异国情调与中国的古典情韵相互映衬，更显贵气尊荣。

为了打造富豪门第、都会城堡的格局，设计师采用了大量对比强烈的色彩和耀眼夸饰的艺术造型，以展现时尚庄园豪邸的大气尺度。设计师运用西方古典工艺的严谨精湛工法，将东方精神注入其中。整体空间氛围传递著西方的浪漫，同时轻诉着东方的曼妙——透过古典与现代装饰艺术的交汇融合，显示出复古、融会、创新与再生的精神。

一踏入室内，触目所及是象征圆满的黑底白圈岗石拼花地板，一路延伸铺满整个中介空间。伞型、拱型的圆弧语汇在空间中展开，炫示着贵族世家的优雅门风，华丽中而有所矜持，打造出庄园豪宅的非凡气度。黑底白圈的地坪延伸到起居室，不銹钢镜面的伞状顶棚，左右对称的黑金花加上金色镀鈦板拱型穿廊

和窗花，使轴线变得方正，也让入口成为正式的迎宾空间。

起居室内，重复的白色框形线板喷漆隐藏了棋牌室和次卧室的门片，同时强调了主卧室的黑色双开门。贴饰线条夸张明快的闪电语汇壁纸，彰显出大户人家圆满中亦有果决独特的个性。

棋牌室入口右侧墙面，胡桃木如精美雕刻的巧克力方块，配上金色编织壁纸，让具有休闲娱乐功能的棋牌室更见低调奢华。

公共空间正是展现业主身份与气度的美学剧场，这座宽广的殿堂界定了业主的不凡品位，也让所有访客的心灵有如经过一场美学的洗礼，充分享受欣赏美丽事物的满足感。风华绝代而不遗世孤立，富丽尊荣而不降格媚俗，流金岁月似乎在空间中隐约流转，构建了本案独特壮阔的庄园豪宅风格。

六角形的浮雕顶棚大胆地布满整个客餐厅，让西方的异国情调与中国的古典情韵相互映衬，更显贵气尊荣。浮雕效果的顶棚延伸至餐厅主墙，强调立体而别具个性的空间感。新古典优雅而不繁复的桌椅造型，强调质感而非画栋雕梁的装修家饰，虽然低调，反而更凸显了整体空间的奢华感。

在卧室中，床头背景上大面积的窗帘，运用间接灯增加层次感，创造出利于休憩身心的优雅氛围。左右对称的更衣空间，黑门框镶嵌不锈钢细边，让男女业主各自拥有仿佛精品店贵宾试衣间的私密舞台。端景墙贴饰着昂贵的孔雀壁纸，呈现独一无二的灿烂华丽质感。

透过设计者巧思铺陈，空间中展现着犹如戏剧般的张力，炫示着贵族世家的优雅门风，华丽中有矜持，打造出庄园豪宅的非凡气度。

Gold Days with Rational Ornaments

The future of design,
Lies in respect for the past,
Ingeniously and intangibly in longing for tomorrow

By Designer, Jeffrey Bilhuber

The real elite mansion does not lie in the common charms of gold and silver lining, but in displaying grand scale and magnificent atmosphere. Hexagonal relief ceiling covers the whole living room in some bold way. The exotic charms of west and classical Chinese sensations bring out the best in each other, presenting magnificent and noble temperament.

In order to create the layout of mansion and metropolitan castle, the designer applies colors of intensive contrast and artistic modeling of brilliant decorations to display the magnificence of fashionable mansion. The designer applies the intricate and rigorous techniques of classical western technique to instill oriental spirits inside. The whole space atmosphere sends out western romance and oriental grace. Through the combination of classical and modern decorative art, antique, integrating, innovative and regenerating spirit is created.

Once you enter the room, what come into eyes are marble block floors representing perfection, extending throughout the whole intermediary space. Arch architectural language of umbrella form spreads out in the space in some clear way, displaying the elegant style of the aristocratic family, with understatement in magnificence, creating the uncommon temperament of the mansion. The terrace extends towards the living room. Umbrella-shaped ceiling of stainless steel mirror surface, symmetrical black gold marble, gold titanium plate arch corridor and window paper cut make the axes become square, while making the entrance become the official reception space.

Inside the living room, the doors of the poker room and the secondary bedroom are hidden behind the repeated white framing molding and paint, while the black double door of the master bedroom is highlighted. The exaggerated and brisk wallpaper of the decorative lining display the consummate and resolute characteristics of rich and influential family.

On the right side of the poker room's entrance, the walnut is like delicate chocolate square, accompanied with gold knitting wallpaper, which makes the poker room of leisure and entertainment functions display and low-key luxury.

Public space is the aesthetic theatre to display the property owner's status and bearing. This broad palace defines the uncommon taste of the property owner, which makes the heart of every visitor seem to experience some aesthetic baptism and fully enjoy the satisfaction of appreciating nice things. This space possesses unprecedented beauty, but does not exist in some secluded way. Gold days seem to flow inside the space, constructing broad and spectacular mansion style.

Hexagonal relief ceiling boldly covers the whole guests' dining hall, making western exotic charms and Chinese classical charms reflect with each other, appearing noble and aristocratic. The ceiling of relief effect extends towards the main wall of the dining hall, stressing on three-dimensional but peculiar space sensations. The desks and chairs are neo-classical elegant and concise style, emphasizing on decorations of texture but no paintings. Although the style is understated, the luxury feel of the whole space is highlighted.

Inside the bedroom, large area of curtains on the background of the bed's head applies indirect lights to add sense of layers, hence creating some elegant atmosphere to rest one's heart. The eudipleural dressing room has black door frame with stainless steel lining, creating host and hostess seem to possess a private stage which is like boutique fitting room for VIPs. The wall with decorations has expensive peacock wallpaper, displaying unparalleled splendor texture.

Through ingenious design of the designer, the space displays tension just like an opera, displaying the elegant atmosphere of rich and influential family, creating uncommon temperament of manor mansion.

远中风华八号楼

Yuanzhong Fenghua, Building No. 8

设计单位：玄武设计
设 计 师：黄书恒、欧阳毅、陈佑如
软装设计：玄武设计
项目地点：上海市静安区
项目面积：200.97 m²
主要材料：雪白银狐大理石、白水晶石材、米洞石、金箔、银箔、水晶、金镜、明镜、图腾雕花板、贝壳板、BISAZZA 陶瓷锦砖、VIVA 瓷砖
摄 影 师：王基守

Design Company: Sherwood Design Group
Designers: Huang Shuheng, Ouyang Yi, Chen Youru
Soft Decoration Design: Sherwood Design
Project Location: Jing'an District in Shanghai
Project Area: 200.97 m²
Major Materials: White Silver Fox Stone, White Crystal Stone, Beige Travertine, Gold Foil, Silver Foil, Crystal, Gold Mirror, Mirror, Totem Carve Plate, Shell Board, BISAZZA Ceramic Mosaic Tile, VIVA Brick
Photographer: Wang Jishou

玄武设计以维多利亚时期古典柔美的设计风格，演绎今日上海精神，以华丽又怀旧的多种元素，创造出优雅、经典、自成一格的新上海风。它的用色大胆绚丽、对比强烈，中性色与褐色、金色结合突出了豪华和大气；它的造型细腻、空间分割精巧、层次丰富、装饰美与自然美完美结合，更是唯美主义的真实体现。

于顶棚和壁面上加入线板、镂空窗花等装饰元素，以婉约线条创造典雅的空间氛围，织品的柔软接口，承载了皇室风格的优雅，图案地板拼花与雕花窗棂相互对应，呈现飞扬时代的复古精神。

穿过走道之后，向左、向右运用不同纹路与质感的壁纸、马赛克拼贴，结合明镜与线板、瓷砖、石材，为每一位居住者打造彰显个人品位的舞台，同时也是让心灵充分休憩的花园，让人在复古的纯净格局中，忘却一日辛劳。

公共空间运用大量的浅米色及白色，搭配淡金、银箔、少数黑色描边与优雅的浅天空蓝，一眼望去，如同跨进魔衣橱里的美丽新境界，隔绝门外尘嚣，蓦然进入皇家御苑之中，为其高贵气势所震慑。

Sherwood Design uses classical and soft Victorian design style to display Shanghai spirit of nowadays and creates elegant and classical new Shanghai style with magnificent and nostalgic elements. Its bold colors display intensive contrast. The combination of neutral, brown and gold colors sets off magnificence. All those are the true representation of aestheticism, meticulous modeling, careful space division, rich layers and the perfect combination of decorative beauty and natural beauty.

The designer adds some decorative elements such as molding, hollow-out paper cut on the ceiling and the wall, creating elegant space atmosphere with graceful lines. The soft interface of fabrics inherits the elegance of royal style. The floor pattern corresponds with the window carvings, displaying the ancient spirit of modern days.

After you walk past the corridor, you can find the wallpaper and mosaic collage of different pattern and texture on both sides. The combination of molding, mirror, ceramic tiles and stones create some stage for the inhabitants to display their personal

taste. While at the same time, this is the garden for people's heart to get complete relaxation. Thus, people would forget about the day's fatigue within this antique and pure space.

The public space makes use of a lot of light beige color and white color, accompanied with light gold, silver foil and elegant light sky azure. When you look around, it is like being among the beautiful new world of magic wardrobe. With the hustle and bustle of mundane world left outside, it is like some imperial garden, and people would get shocked by its noble momentum.

深圳·宝安中洲中央公园 3-A01户型样板房

Shenzhen · Baoan Centralcon Central Park 3-A01 Showroom

设计单位：KSL 设计事务所
设 计 师：林冠成、温旭武、马海泽
项目地点：广东省深圳市宝安区
项目面积：176 m^2
主要材料：皮革、特殊玻璃、灰茶镜、珍珠米黄大理石、木地板

Design Company: KSL DESIGN(HK)LTD.
Designers: Lin Guancheng, Wen Xuwu, Ma Huize
Project Location: Bao'an District, Shenzhen, Guangdong Province
Project Area: 176 m^2
Major Materials: Leather, Special Glass, Grey Tawny Glass, Pearl Beige Marble, Wood Flooring

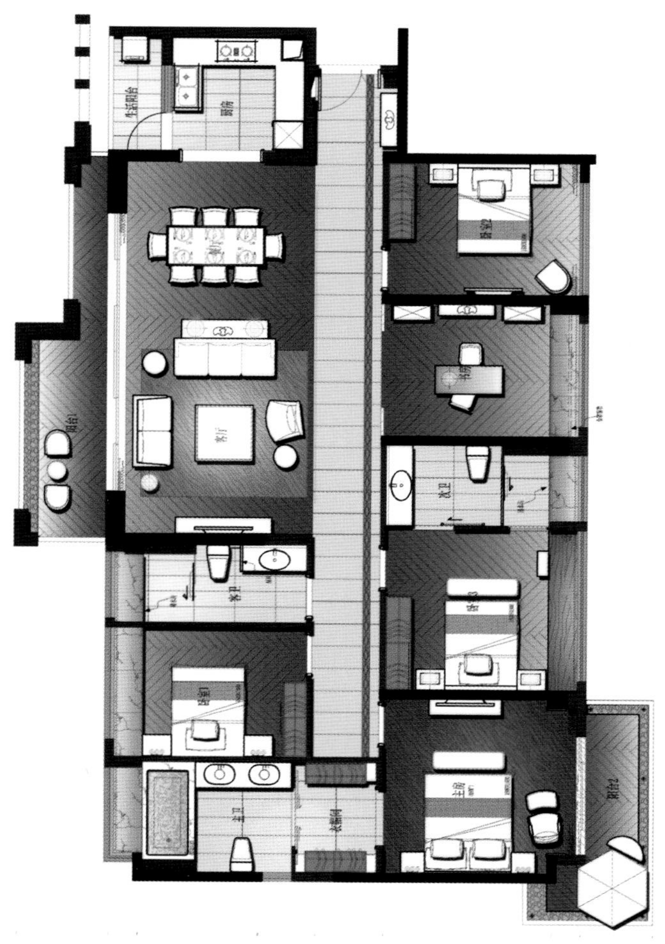

设计师采用了现代的设计手法，将法式特有的典雅融入现代的简洁，从而迸发出另一种独特的浪漫风情。层层递进的色彩与造型将空间延伸，让人可以感受到空间中蕴涵的大家风范。长长的走廊将所有的功能空间串联在一起，像一条流淌的河水一般让各个空间相互联系、充满活力。整体色调淡雅、稳重且充满品位，鲜艳的红玫瑰在室内默默地绽放，仿佛带我们回到了过去的时光，将记忆点缀。

家居环境中摆放着精致的灯具、通透的玻璃器皿、简洁的布艺，无意间为空间增添了更多的浪漫元素。设计师致力于营造一个简约、大气、充满浪漫而又具有轻松感的家居空间，在淡淡的玫瑰香气与素雅的装饰中，自有一种出众的现代法式气氛环绕于身旁，让人沉浸于空间所营造的美好氛围中，心生欢喜。

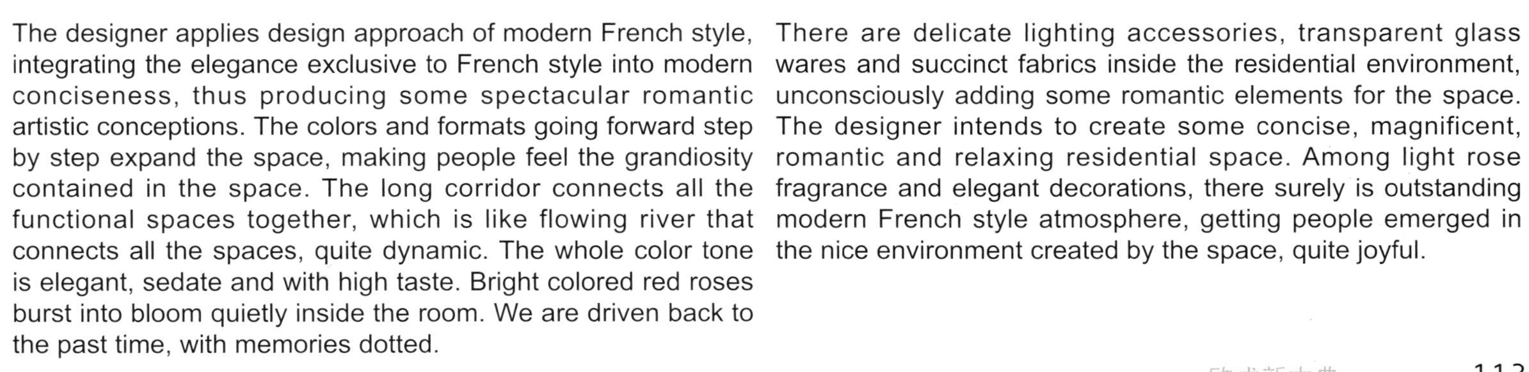

The designer applies design approach of modern French style, integrating the elegance exclusive to French style into modern conciseness, thus producing some spectacular romantic artistic conceptions. The colors and formats going forward step by step expand the space, making people feel the grandiosity contained in the space. The long corridor connects all the functional spaces together, which is like flowing river that connects all the spaces, quite dynamic. The whole color tone is elegant, sedate and with high taste. Bright colored red roses burst into bloom quietly inside the room. We are driven back to the past time, with memories dotted.

There are delicate lighting accessories, transparent glass wares and succinct fabrics inside the residential environment, unconsciously adding some romantic elements for the space. The designer intends to create some concise, magnificent, romantic and relaxing residential space. Among light rose fragrance and elegant decorations, there surely is outstanding modern French style atmosphere, getting people emerged in the nice environment created by the space, quite joyful.

与有形处赏无形之美

Invisible Beauty in the Visible Location

设计单位：上海元尚装饰设计有限公司
设 计 师：孙扬元
项目地点：浙江省杭州市西溪望庄
项目面积：309 m^2
主要材料：乳胶漆、壁纸、大理石、木地板、仿古砖、地毯

Design Company: Shanghai Yuanshang Decorative Design Co., Ltd.
Designer: Sun Yangyuan
Project Location: Hangzhou in Zhejiang Province
Project Area: 309 m^2
Major Materials: Emulsion Paint, Wallpaper, Marble, Wood Floor, Antique Brick, Carpet

文艺复兴时期最伟大的建筑师之一安德烈·帕拉迪奥（Andrea Palladio）曾经说过："最优美、最规则的形状是圆形和正方形，其他形状都是由它们主导出来的。"本案传承的正是这个理念，并以此为基础，为现代住宅注入了古典的内在气质。

大气的空间规划

16 世纪，帕拉迪奥曾为梵蒂冈的神职人员设计过一套"圆厅别墅"，房子的外形轮廓是一个十字，而十字正中间的部分则是一个圆厅。本案正是借鉴了"十字圆厅"的空间设计。在公共区域呈现出明显的十字走向，同时通过一个过渡空间及简明的纵横动线，将会客区与私人区明显地区分开来。本案在运用"十字圆厅"这一概念的同时，在技法上破立结合，更多地传达出这一理念的内在气质。

舒适的起居享受

本案的三间朝南卧室正对着两侧栽有法国梧桐的林荫大道，与室内华贵的法式情调相互交融，极为浪漫。主卧室运用了星级酒店总统套房的设计概念，以米白色为基调，穿插金色、黑色、酒红色，呈现出静谧奢华的气氛。主卧室的面积超过 80 m^2，有近 8m 的开间，两扇落地飘窗使

业主将窗外的海棠枝条、黄色鸢尾尽收眼底，晴时阳光洒满一室，雨时银丝轻坠，斜倚床榻就能欣赏江南朦胧的烟雨，真有一番“星沉海底当窗见，雨过河源隔座看”的意境。

多元化材质的运用

本案在材质的运用上相当多元化。餐厅墙面上的镜面材质，使空间在视觉上得到了延展。改造后的客房空间不够宽敞，设计师就在墙上挂置了镜子，同时利用梳妆台的镜面形成双倍拓展效果。客厅里，玻璃拼接的透视茶几最大程度减弱了家具的体积感，从而使视觉得到舒展。与此相对，过道入口处的会客区同样使用了玻璃门，保持了空间在视觉上的完整性。主卧室里，米色石板拼接背景墙上的大幅镜面，搭配烤漆镜面边柜，则是为了遥相呼应，锦上添花。

设计师语录

这套居室以新古典风格作为主导，整个空间以法式的情调为主导，同时借鉴了一些酒店设计理念。不同的功能空间散发着不同的艺术气息，每个区域都有自己的独特内涵，但设计思路是一致的——旨在寻找有形的美。从入口开始，利用一些古典设计手法去演绎这个浪漫之家。整体色调以米色为主，客厅里采用了玻璃、镜面等现代材质，以营造空间的大气感。

Andrea Palladio, the greatest architect of Renaissance period, once said, “The most elegant and most regular shapes are round and square, and all other shapes derive from them.” What this project inherits is this concept. And based on this, modern residence is instilled with classical inherent temperament.

Magnificent Space Planning

In the 16th century, Palladio designed Villa Rotonda for clergymen of Vatican. This villa has cross-shaped outline with a round hall in the middle of the cross. And this project takes the example of the space design of Villa Rotonda. The public space has clear cross layout. And through some transition space and concise vertical and horizontal lines, the meeting area and private area are clearly separated from each other. While making use of the concept of Villa Rotonda, this project carries out flexible application in the techniques, conveying the inner temperament of this concept in some further way.

Enjoyment of Comfortable Residence

The three south-facing bedrooms of this project face the boulevard with cotonier on both sides, integrating perfectly with the interior magnificent French charms, quite romantic. The master bedroom applies the design concept of presidential suite in star hotel. With creamy white as the tone color, interlaced with gold, black and burgundy, the design unfolds quiet and magnificent atmosphere. The area of the master bedroom is over 80m^2, with an open-style room with 8m in length. Two French windows make it possible for the property owners to enjoy the branches of begonia and yellow fleur-de-lis. When it is sunny, sunshine spreads inside the whole space. When it is rainy, rainwater drops gracefully outside the window, thus the property owner can enjoy the misty rain while lying on the bed.

The Application of Multiple Materials

The application of materials for this project is quite diversified. The mirror surface on the dining hall’s wall visually expands the space. The guest room space after transformation is that capacious, thus the designer hangs a mirror on the wall and achieves expansive effect with mirror of the dresser. Inside the living room, the transparent tea table of glass minimizes the scale of the furniture, thus creating comforting visual effects. Compared with that, the reception area at the entrance also applies glass door, maintaining visual integrity of the space. Inside the master bedroom, the large mirror surface on the background wall of the beige stone plate is accompanied with side cupboard, accomplishing some better overall effect.

Remarks of the Designer

The whole residence is guided by neo-classical style and French emotional appeals, while having the design concepts of some hotels for reference. Different functional spaces send out different artistic atmosphere. Every region has its peculiar connotations. But the design trains of thoughts are consistent – aspiring to search for the tangible beauty. From the entrance, the designer makes use of some classical design approaches to create some romantic home. The whole color tone centers on beige. The living room applies modern materials such as glass and mirror to create some magnificent atmosphere inside the space.

九龙仓·雍锦汇示范单位

Jiulongcang• Yongjinhui Sample Flat

设计单位：香港方黄建筑师事务所
设 计 师：方峻
项目地点：四川省成都市
项目面积：200 m^2

Design Company: Hong Kong Fang Huang Architects Studio
Designer: Fang Jun
Project Location: Chengdu in Sichuan Province
Project Area: 200 m^2

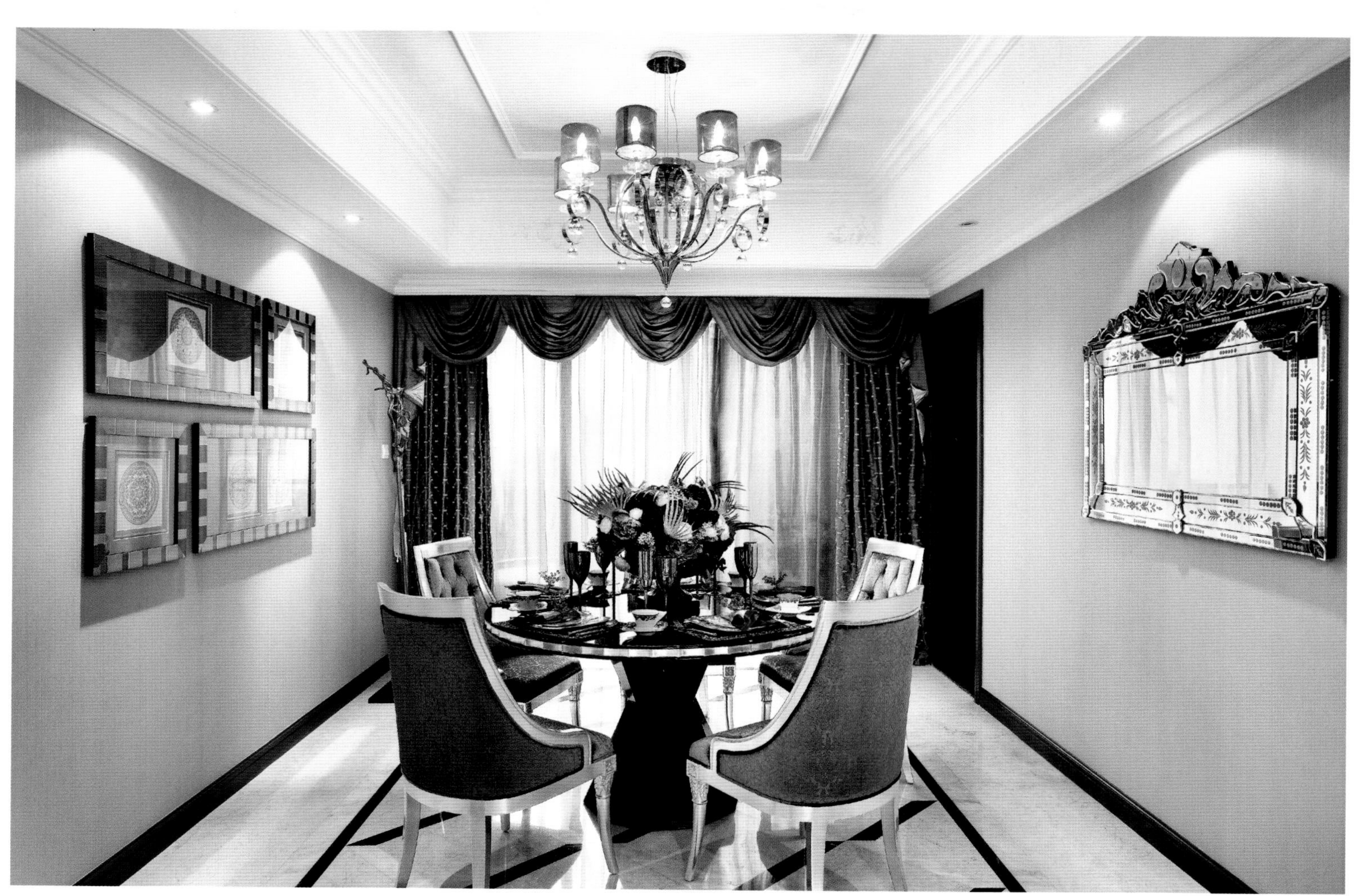

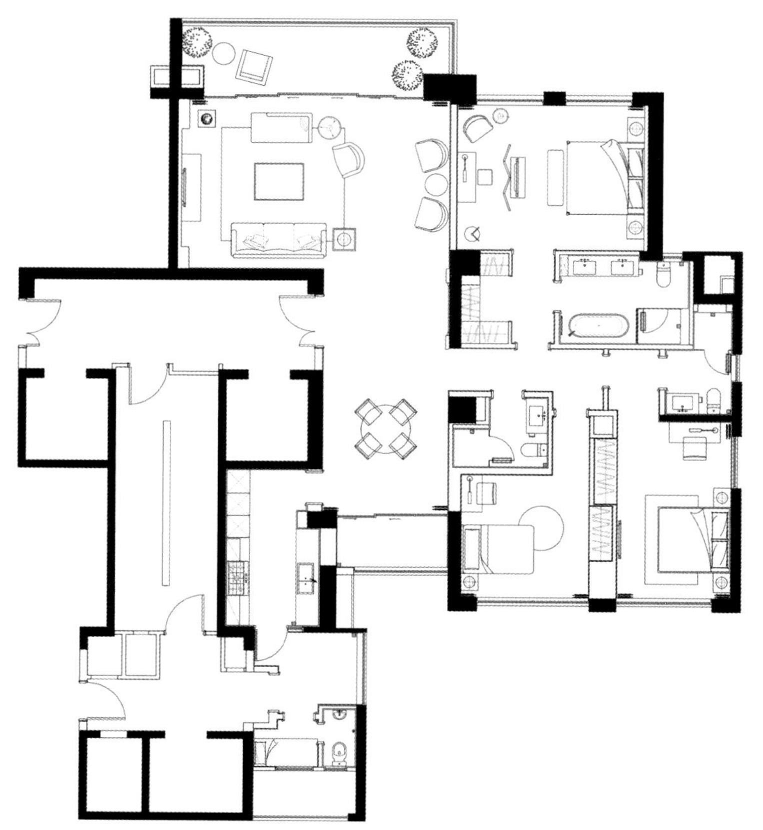

本案为新古典主义风格，奢华的洛可可风格充斥于每个角落之中。设计师以棕色、黑色等色彩来奠定空间华贵、沉稳的基调，空间内到处摆放金色、银色制品，增强了空间的奢华氛围。

入口处并未做复杂的装饰，一个摆台、几件金银制品和两个黑色天鹅工艺品，却在无形中提升了业主的品位。棕色绒布地毯、蓝底大面积的图腾地毯、洛可可风格的家具围合而成的休闲区，奢华气息扑面而来。

This project is Neo-Classical style, with luxurious Rococo style in every corner of the space. The designer applies brown and black colors to set the noble and calm tone of the space. There are gold and silver color accessories everywhere in the space, enhancing the luxurious atmosphere of the space.

There are no complicated decorations at the entrance. A stand, several gold and silver accessories and two black swan artistic objects virtually promote the taste of the property owner. The leisure space enclosed by brown lint carpet, totem carpet of large area blue color and furniture of Rococo style, producing magnificent luxurious atmosphere.

万豪国际

Marriott International

设计单位：武汉郑一鸣室内建筑设计
设 计 师：郑一鸣
软装设计：吴锦文
项目面积：300 m²
主要材料：墙板、艺术壁纸、大理石、陶瓷锦砖、实木地板、欧式家具

Design Company: Wuhan Zheng Yiming Interior Architectural Design
Designer: Zheng Yiming
Soft Decoration Designer: Wu Jinwen
Project Area: 300 m²
Major Materials: Wallboard, Artistic Wallpaper, Marble, Glass Ceramic Mosaic Tile, Solid Wood Floor, European Furniture

本案为武汉临湖豪宅，地处繁华的市中心，住宅南北通透，左右临湖。建筑原本的结构较为开阔，设计师通过隔断将空间划分为多个层次，使空间更加丰富。门厅采用墙板对称造型，点缀水晶壁灯，与装饰镜交相辉映，使门厅的气氛唯美、华丽。

客厅中，雕花造型作为沙发的背景，一方面将原本的异形结构变得方正起来，同时能作为沙发的靠背，造型通透、细腻。背景墙为淡蓝色的樱花图案，左右为石材欧式墙板，既端庄又与整体的造型相呼应。淡蓝色的樱花图案也体现出女主人娟秀、时尚的气质。门厅的左边有两扇对称的玻璃推拉门，进去后便是娱乐室与保姆房，不同空间相互独立，在娱乐的同时又不会影响到其他空间。

This is a lakeside mansion located in Wuhan, at the prosperous urban center. This mansion is transparent in the north-south direction and has lake views on the left and the right side. The original structure of the building is kind of broad. Through partitions, the designer separates the space into several layers to enrich the space. The door hallway applies symmetrical format of door plate, accommodated with crystal. The hallway and the decorative mirror add beauty and radiance to each other, making the atmosphere of the space aesthetic and magnificent.

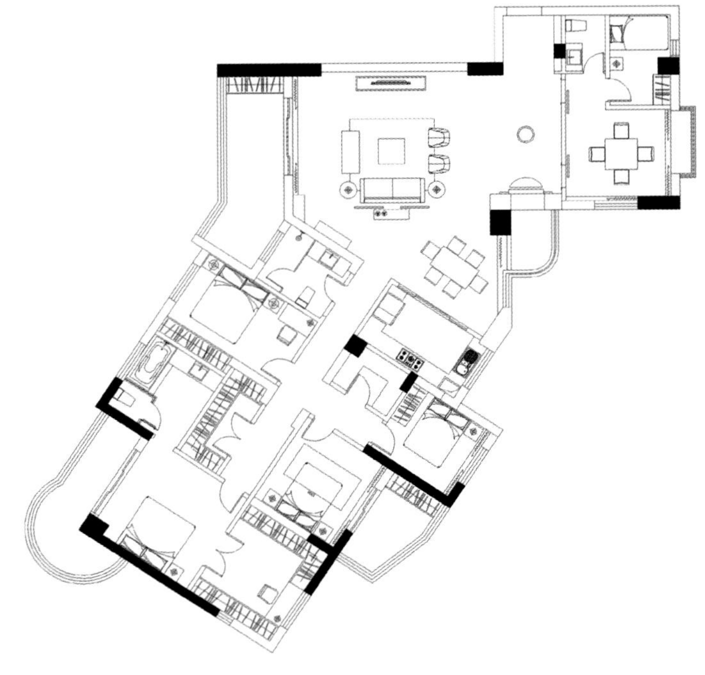

For the living room, the carving pattern has sofa as the background. On one hand, the carving makes the original structure appear square, on the other hand, it can be used as the back for the sofa, with transparent and refined style. The background wall has light blue cherry blossom graphics, on both sides European stone wall plates, appearing modest and in harmony with the model of the whole space. The light blue cherry blossom graphics represent the graceful and fashionable temperament of the hostess. On the left side of the hallway is two symmetrical glass sliding doors. Inside are the entertainment room and the nanny's room. Different spaces are mutually independent, not influencing other spaces while providing entertainment functions.

大东城二期样板房

Phase 2 Show Flat, Dadongcheng Property

设计单位：J&D 设计事务所
设 计 师：潘冬东
项目面积：124 m²
主要材料：大花白云石、法国灰云石、壁纸、金属陶瓷锦砖

Design Company: J&D Design Studio
Designer: Jason Poon
Project Area: 124 m²
Major Materials: Italian White Marble, French Grey Marble, Wallpaper, Metal Mosaic Tile

淋浴
卫生间
儿童房
书房
餐厅
厨房
浴缸
装饰画
装饰柜
客厅
娱乐室
壁炉
TV
老人房
TV
主卧室
TV
阳台

本案为龙岗坪山新区大东城二期的样板房，整体设计以营造清新浪漫的薰衣草氛围作为主导，通过运用古典贵气的欧式元素结合花纹壁纸，给人奢华中散发诗情画意的感受。本案的功能分区明确，流线顺畅，设计师将开放区域和私密区域有意识地划分开来，使其相互联系又互不干扰。

客厅的设计以浅色调为主，白云石的铺地让空间显得宽敞、明亮，沙发背景墙上细腻的石膏装饰纹样搭配着柔软的紫色布艺，凸显出家居环境的干净、纯洁。主卧的设计也延续了这一风格，浅紫色的布艺像是来自于一个遥远的梦境，将人笼罩在一种安静的氛围中，空气中仿佛还弥漫着薰衣草淡淡的香气，让人在安心的气息中可以享受到静谧的睡眠环境。不同肌理材质的搭配，创造出不同的空间体验。书房、餐厅、儿童房的设计从业主的角度出发，所使用的家具和灯光的搭配突出了时尚浪漫的主题风格。

This project is a show flat with design guided by fresh and romantic lavender. With the application of classical and aristocratic European elements and decorative pattern wallpaper, the design leaves people with poetic impression within this luxury. This project has distinct functional area divisions with fluent lines. The designer consciously separates the open area from the private area, to make them connect but not disturb each other.

The design of the living room focuses on light color tone. White marble pavement makes the space broad and bright. The refined gypsum decorative pattern on the background wall of the sofa goes with the soft purple cloth, highlighting the clarity and purity of the residential environment. The design of the master bedroom continues this style. Light purple cloth is like from some remote dreamland, enveloping people in some quiet atmosphere. There seems to be light fragrance of lavender in the space, making people enjoy tranquil sleeping environment in this quiet atmosphere. The collocation of materials of different texture creates some different space experience. The design of the study, dining hall and children's room starts from the point of the property owner. The combination of furniture and lights highlights the fashionable and romantic theme and style.

法式血统的优雅年代

Elegant Age of French Style

设计单位：大匀国际空间设计
设 计 师：林宪政、陈雯婧
软装设计：太舍馆贸易有限公司
项目地点：浙江省宁波市
项目面积：137.3 m²
开 发 商：宁波镇海维科房地产开发有限公司
主要材料：白色喷漆、表布、柚木

Design Company: Symmetry Design Center Co.,Ltd
Designers: Lin Xianzheng,Jessie Chen
Soft Decoration Designer: MOGA DECO Co., Ltd.
Project Location: Ningbo in Zhejiang Province
Project Area: 137.3 m²
Developer: Ningbo Zhenhai Weike Real Estate Development Co., Ltd.
Major Materials: White Paint, Cloth, Teak

MASTER BATH ROOM
-20
BED ROOM
±0
MASTER BED ROOM
±0
TV
TV
BATH ROOM
-20
BALCONY
-20
STUDY ROOM
±0
TV
LIVING ROOM
±0
KITCHEN
-20
DINING ROOM
±0
MAID'S ROOM
±0
REF

迷情法兰西

白色镶金色系的门框、窗户、桌子和浅粉绿色的壁纸调配出迷人的法式情调，素雅的小碎花椅面、白色器皿、低彩度的空间线条结合地面的几何图形，赋予空间优雅而浪漫的法兰西女性气质。

法式风格讲究将空间融入自然中，在设计上追求心灵的自然回归感，给人一种扑面而来的清新气息。所以，开放式的空间结构给予了各个功能空间相互交流的机会，整体流线的贯通也使得业主一家生活得更加惬意。在空间的装饰层面上，设计师并未过多地采用华丽的陈设品，而是放置了随处可见的花卉和绿色植物、雕刻精细的家具、田园气息的布艺……所有的一切从整体上营造出一种贵族浪漫之气，同时融合了小清新的流行都市概念。在任何一个角落，都能体会到业主悠然自得的生活形态和阳光般明媚的心情。

Obsessed with French Style

The extensive white is accompanied by golden door frames, windows, tables and light pale-green walls, introducing the charming French style. The refreshing flower-patterned seat covers, white containers and plainly-colored space lines, combined with geometric figures of the ground, endow the space with the elegant and romantic French feminine style.

According to the French style, the space should be placed in nature. The design emphasizes people's spiritual return to nature and the refreshing atmosphere. Therefore, designers decorate the open space structure with commonly-seen flowers, green plants and sophisticated furniture, all of which create an aristocratic and romantic style and convey the popular urban concept focusing on the refreshing quality. The leisurely life and good mood of the owner can be found in each corner.

BOBO城样板房

Show Flat of BOBO City

设计单位：宁波十杰装饰工程有限公司
设 计 师：卓咪咪
软装设计：上锦艺术陈设
项目面积：150 m^2
主要材料：大理石、壁纸、花梨木地板
摄 影 师：刘鹰
撰　　文：刘思恩

Design Company: Ningbo Shijie Decoration and Engineering Co., Ltd.
Designer: Zhuo Mimi
Soft Decoration Designer: Shangjin Artistic Layout
Project Area: 150 m^2
Major Materials: Marble, Wallpaper, Rosewood Floor
Photographer: Liu Ying
Composer: Liu Si'en

一生一宅

风光秀美的城市，样式简约的建筑，散发着欧洲的浪漫气息，不时透出的烂漫花丛，云涛海浪中，渐行渐远。碧波荡漾的水中点缀着青荷红莲，葱茏中掩盖着清晰明亮的屋檐，一生的住宅选择在此微笑。

本案的住宅宣言是一生一宅。欧式新古典是户型所采用的设计风格，整体为偏浅色系的淡雅、舒适。宽敞大气的电视背景墙、家具、饰品的选择以及拼铺地砖造型展现出一个雍容大气、优雅别致的新古典风格客厅。

整体空间的浅色系壁纸，大气中又透出细腻，设计师在细节的把握上可见一斑。餐厅将客厅的华贵进一步延续，水景烛台下端起酒杯享受独特的人生。过道的尽头设计了一组端景台，银色马头的雕塑吸引人驻足欣赏。以深色为主色调的书房端庄、雅致、高贵，同时体现出业主的沉稳之气。宽敞的空间发挥了它的优势，使流畅、开阔的公私空间以最自然的形态表现出来，书桌上的冷色调台灯和顶棚上的水晶吊灯亦协调得恰到好处。

清晨透过树叶的斑驳的阳光倾洒下来，落在柔软、优雅的寝具上。深红色的软包，让空间柔美许多——女主人钟爱红色，屋内的花艺选择、饰品摆件均为红色系。轻盈的窗帘，柔软的地毯，洁净的墙壁，华美而尊贵。男孩房以航海为主调——小男孩梦想成为航海员——以白色、蓝色为主的海洋色系，井然有序的摆件，均突出了男孩对大海的向往。

纯净的白色，总是让人心旷神怡地闭目养神，深陷其中。暮色黄昏，五彩斑斓的船只在扬帆待行，等待着追求纯白梦想的人们踏上一生中绚丽的五彩追梦之旅。

A Mansion for A Lifetime

A city of nice sceneries and building of concise outlook send out European romantic atmosphere. Bright flower shrubs here and there fade away gradually in the clouds and sea waves. The water of ripples is dotted with green lotus leaves and red lotus. A clear and bright eave is hidden in the verdant plants. A mansion for a lifetime is smiling here.

The manifesto of this project is A Mansion for A Lifetime. The flat here applies neo-classical European design style. The whole project is elegant and comfortable light color tones. Spacious and generous TV background wall, furniture, ornaments and floor tile display noble, magnificent, elegant and spectacular neo-classical living room.

For the light color wallpaper of the whole space, there is delicateness in the magnificence. The dining hall continues the magnificence of the living room. You can enjoy your own life with a cup of wine in hand under the waterscape candlestick. There is a platform with ornaments at the end of the corridor. Silver horse's head sculpture invites people to stand and appreciate. The study with dark color as the tone color is dignified, elegant and noble, while displaying the property owner's sedate temperament. Broad space brings full play to its strengths, making fluent and open space present itself in the most natural way. Cold color lamps on the desk and the crystal droplights on the ceiling coordinate well with each other.

In the morning, mottled sunshine shines through the leaves and scatters on the soft and graceful bedding accessories. Dark red soft rolls make the space appear much softer. The hostess is fond of red colors and the interior floriculture and ornaments are all red colors. Light curtains, soft carpet and clear walls are splendid and aristocratic. The little boy dreams to become a navigator. Thus the boy's room centers on navigation as the tone and selects ocean color schemes such as white and blue. Well-regulated ornaments highlight the little boy's longing for the ocean.

Pure white would always help people rest to attain mental composure in some relaxed and happy way. At dusk, colorful ships are waiting to set sail, waiting for people pursuing pure white dreams start on the journey of chasing dreams.

福清金辉华府样板房

Jinhui Mansion Show Flat

设计单位：福州北岸装饰设计有限公司
设 计 师：肖鸣
项目面积：160 m²
主要材料：阿曼米黄石材、泛美木地板
摄 影 师：周跃东

Design Company: Fuzhou North Bank Decorative Design Co., Ltd.
Designer: Xiao Ming
Project Area: 160 m^2
Major Materials: Arman Beige Stone, Pan American Wood Floor
Photographer: Zhou Yuedong

每一个家都有着耐人寻味的故事——因为每个人心中都有许多挥之不去的情结，将这些情结在家中以自己钟爱的形式所展现，它带来的不仅是视觉上的体验，更是触动心灵的一个过程。走进福清金辉华府样板房，开放式的空间呈现出令人踏实的包容感，在与空间的对话中很容易产生情感的共鸣。欧式风情是设计师给予这个空间的视觉标准，但设计师并没有凭借直观意识去堆砌它，而是使用灵感进行再创造。

在客厅、餐厅、书房、卧室这些功能区中，时尚与古典的奇妙共处让人感受着一种气派的名门风范。新古典主义风格的家具以及工整、严谨的布局，将欧式情韵展现无余，错落有致的体量感从不同角度营造出丰富的空间形象。置身其中，精致、厚重的生活之感油然而生，低调奢华的格调处处弥漫。

除了结构与陈设之外，光线的运用也是不能忽略的空间细节，它是让空间鲜活起来的妙法。本套样板房里的主光源与点光源共同调节着室内的光线，它们以不同姿态相互呼应，让人联想起灯光辉映的舞台，一下子活络了空间的每个角落。渐渐地，设计师让古典奢华的概念好似一部写满艺术的剧本，任人们理性地翻开，而后感性地阅读。

Every family would have their stories affording for thoughts -- as every one would have some haunting complex in heart. These complex can be displayed inside the space in some way that the property owner is fond of, which presents not only visual experiences, but also a psychological process. This open space displays some steadfast feel of tolerance. People would easily have some emotional resonance through the dialogue with the space. European charm is the visual standard the the designer wants to give this space. The designer does not want to pile up these elements through his own intuitive consciousness, but recreating the space with inspirations.

For the functional spaces such as the living room, dining room, study and the bedroom. The intriguing coexistence of fashion and classics makes people sense some gorgeous elite style. The furniture of Neo-Classical style is orderly and rigid in the layout, bringing out the European emotional appeals to the extreme. The well-proportioned scale of the space profound some abundant space image from different angles. Being inside the space, people would perceive the delicateness and heaviness of life here. Low-key luxury tone is everywhere.

Apart from the structure and layout, the application of lights is a space detail that people can not neglect. It is what activates the whole space. The major light source and the spot light source together adjust the interior lights, which correspond with each other in different postures. This remind people of

the stage with shining lights, making every corner of the space appear dynamic. Gradually, the designer makes the classical luxurious concept like some script full of art. People can open the book rationally and then read through sensibly.

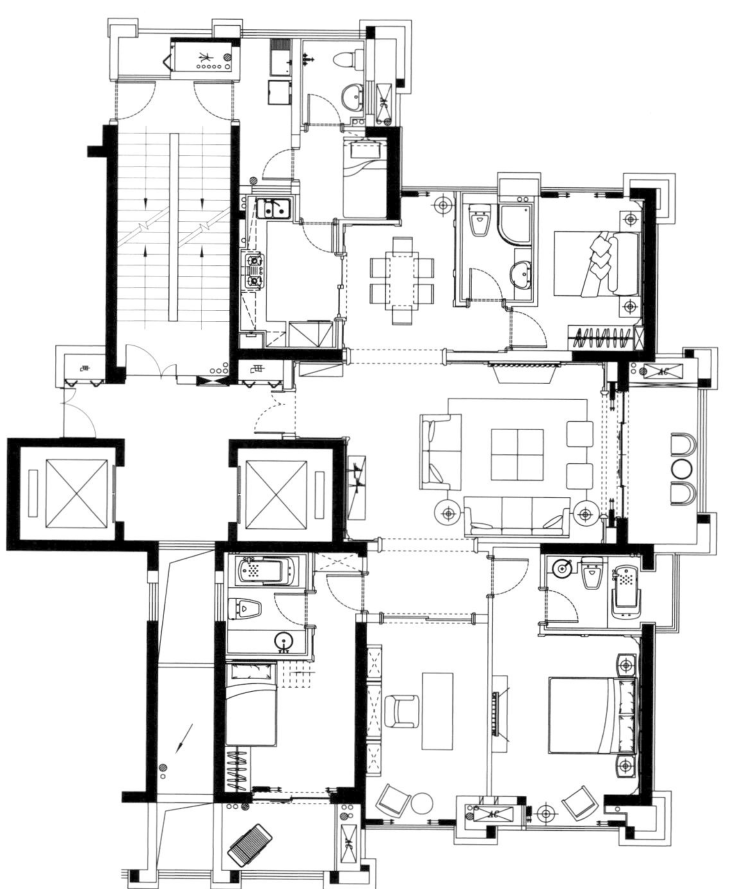

福州万科任先生豪宅

Fuzhou Vanke Mr. Ren's Mansion

设计单位：尚堂空间设计
设 计 师：李风
项目面积：150 m^2
项目地点：福建省福州市新店万科一期
摄 影 师：李琳玉

Design Company: Shangtang Space Design
Designer: Li Feng
Project Area: 150 m^2
Project Location: Vanke Phase 1, Fuzhou in Fujian Province
Photographer: Li Linyu

欧式古典风格一直以来是成功人士的首选，但是，传统的欧式古典风格由于装饰过于繁复、豪华，在造型上又过于古板，已经无法适应于现代人们的需求。因此，设计师采用欧式新古典风格，摒弃了复杂的肌理和装饰，简化了线条，为业主打造出一种独具特色的欧式风情。

本案在传统的欧式中融入了现代时尚的元素，利用黑白灰的搭配、夸张的抽象画和天然纹理的白色大理石营造出雍容淡雅的韵律感，而细节处刻意雕刻而成的线条，更有一种精雕细琢的品位。客厅的装饰、家具的雕花、吊灯上精美的金属镶嵌花纹，做工极为精细，将欧式风格的精髓完美展现，异域色彩颇为浓郁。灰色玻璃及透光板装饰的应用，既带着欧式的典雅，又有着时尚的气息。

European classical style has always been the preference of successful people. But due to the fact that European classical styles are much too complicated and luxurious in the decorations and much too old-fashioned in the modeling, it can no longer meet with the requirements of modern people. Thus, the designer applies European neo-classical style and forsakes much too complicated texture and decorations. With simplified lines, the designer creates some European atmosphere with peculiar characteristics for the property owners.

This project integrates some modern and fashionable elements in the

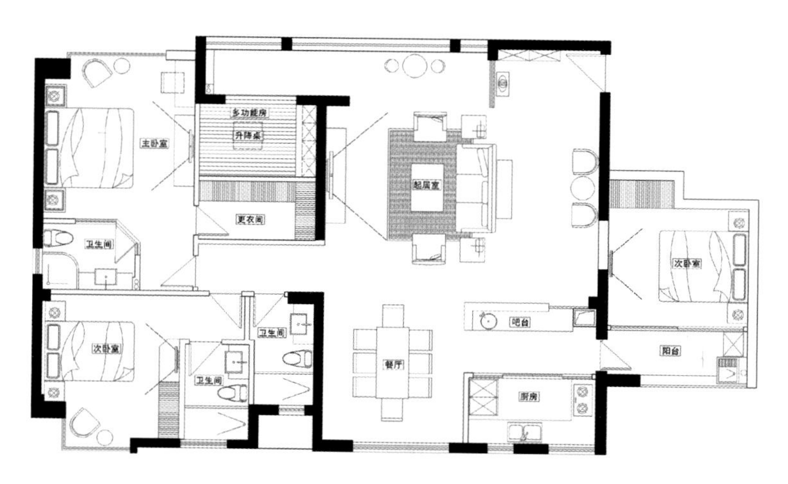

traditional European style and creates some noble and elegant charms in the collocation of black, white and grey, exaggerated abstract painting and white marble of natural texture. And the lines in the detailed parts with deliberation have some refined taste. The decorations in the living room, carvings of the furniture and exquisite metal carving of the droplights are all quite meticulous in the craftsmanship. The essence of European style is presented in some perfect way, with intensive exotic colors. The application of grey glass and light panels not only possesses some European elegance, but also displays some fashionable atmosphere.

六都国际C户型样板房

Sixth City Show Flat,Flat Type C

设计单位：鸿扬集团 · 陈志斌设计事务所
主设计师：陈志斌
参与设计：谢琦
项目地点：湖南省长沙市
项目面积：162 m^2
主要材料：美国米黄大理石、灰镜、黑钢、壁纸、皮质软包、硬包
摄 影 师：管盼星

Design Company: Chen Zhinbin Design Studio of Hirun Group
Chief Designer: Chen Zhibin
Associate Designer: Xie Qi
Project Location: Changsha in Hunan Province
Project Area: 162 m^2
Major Materials: American Beige Marble, Grey Mirror, Black Steel, Wallpaper, Leather Soft Roll, Hard Roll
Photographer: Guan Panxing

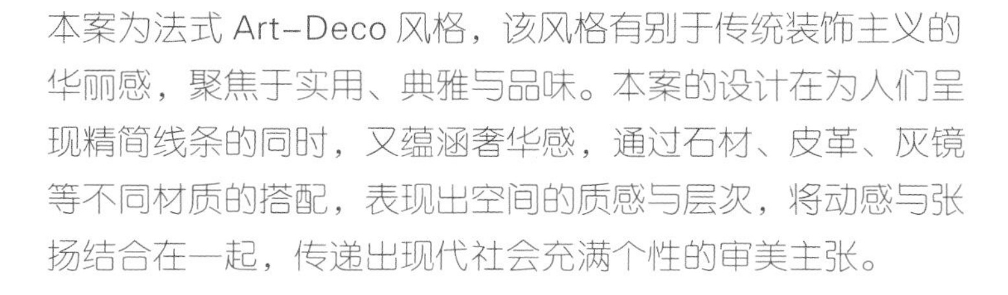

本案为法式 Art-Deco 风格，该风格有别于传统装饰主义的华丽感，聚焦于实用、典雅与品味。本案的设计在为人们呈现精简线条的同时，又蕴涵奢华感，通过石材、皮革、灰镜等不同材质的搭配，表现出空间的质感与层次，将动感与张扬结合在一起，传递出现代社会充满个性的审美主张。

本案融多种风情于一体，既激发了业主对生活的灵感，又使空间显得自由而高贵。在这里，每个功能空间都有其独特的个性，设计师通过在细节处元素的重复使用来营造整体感，使空间风格和谐统一。设计师凭借潮流的生活方式、前卫的设计理念、贴心的设计服务，为业主打造出一个专属的法式奢华家居环境。

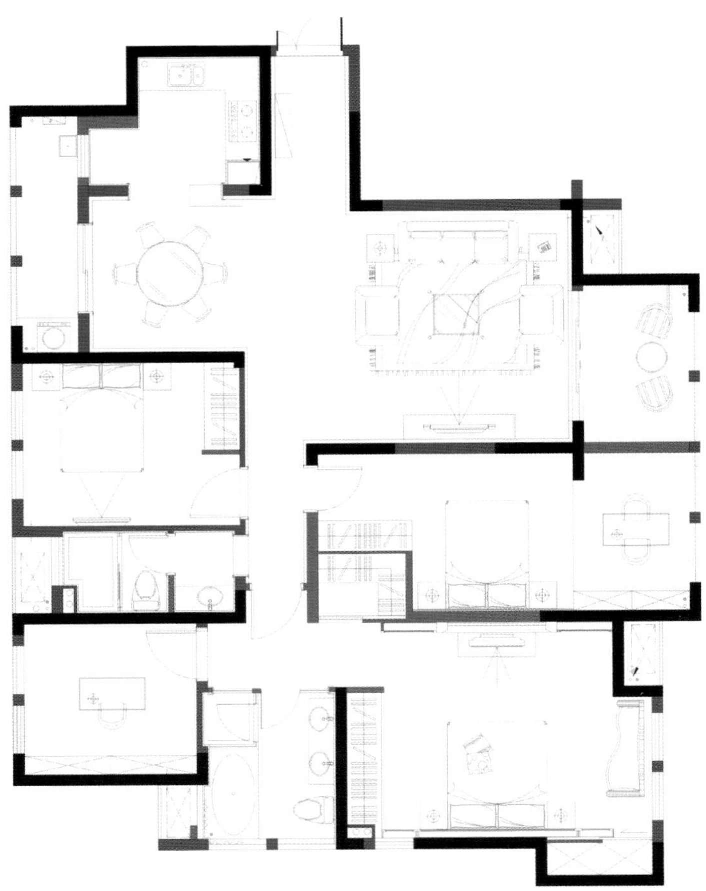

LIFEDBO

This project is French Art-Deco style, which is different from the magnificent traditional decoration, focusing on practicality, elegance and taste. While presenting simplified lines, the design of this project possesses some luxury feel and displays the texture and layers of space through the combination of different materials, such as stone, leather, grey mirror, etc., integrating dynamic and openness, transmitting the individual aesthetic proposition of modern society.

This project integrates multiple amorous feelings which not only inspires property owner's inspirations for life, but also makes the space appear free and noble. Here, every functional space has its peculiar characteristics. The designer tries to create whole feeling through repeated application of detail elements, thus making the space style integrated and harmonious. The designer creates some exclusive and luxurious French style residential environment depending on trendy lifestyle, vanguard design concepts and intimate design services.

名城国际

Mingcheng International

设计单位：福建品川装饰设计工程有限公司
设 计 师：林新闻、俞燕琴
项目地点：福建省福州市
项目面积：160 m²
主要材料：闪电米黄大理石、白玉兰大理石、木面墙裙、壁纸
摄 影 师：周跃东

Design Company: Fujian Pinchuan Decorative Design and Engineering Co., Ltd.
Designers: Lin Xinwen, Yu Yanqin
Project Location: Fuzhou in Fujian Province
Project Area: 160 m²
Major Materials: Beige Marble, White Magnolia Marble, Wood Wainscot, Wallpaper
Photographer: Zhou Yuedong

在这套样板房中，设计师利用白色展开视觉空间的延伸及穿透，并运用空间的层次变化、暖色系元素的搭配和水晶灯光的烘托，创造出“金色茶糜”的丰富空间。设计师将金属的光泽、玻璃水晶的通透、石材的厚实、真皮的质感和布艺的温暖通过各种结合突显出来，在金色灯光的映衬下，空间拥有了不同寻常的美感与质地。空间以暖色系为主旋律，镜面、石材、金属等反光材料的运用强化了灯光色调的晕染，保证了空间的通透感和主题色彩的一致性。整体的金色氛围并没有使空间俗气，反而在带有浓烈欧式色彩的典雅富丽中，将空间表现得华而不俗。

空间的设计颇具浪漫气息。抛开朦胧的金色所营造的迷离般的西洋情调不说，单是家具、灯具的造型与弧度都弥漫着浓浓的法式古典的艺术气息。背景墙的色调以奶白色和棕黄色为主，墙体线条明朗，为软装提供了充分的发挥余地。沙发、茶几、餐桌椅以其独有的欧式宫廷造型和金属雕花工艺吸引着观者的眼球，对称的摆设也流露出古典欧式风格的空间美。欧式风格对于金属和水晶灯的要求较高，恰到好处的灯光来源于细节的打造和对质感的追求。灯具在本案的传统照明功能并没有弱化，美化环境和装饰室内空间的功能却在加强。

For this show flat, the designer uses white color to carry out the extension and transparency of visual space and creates abundant gold space with the layer variation of metal, collocation of warm color elements and crystal lights. Through various combinations, the designer highlights the luster of metal, transparency of glass crystal, heaviness of stone, texture of real leather and warmth of cloth. Set off by gold lights, the space displays some different aesthetic beauty and texture. The space centers on warm color. The application of reflecting materials such as mirror surface, stone, metal, etc., strengthens the color tone of lights, guaranteeing the transparency of space and consistency of theme colors. The whole gold color atmosphere does not make the space appear vulgar at all, but make it display some peculiar and magnificent atmosphere with the intensive European colors.

The space design possesses some strong romantic atmosphere. Apart from the blurred western emotional appeals created by gold colors, all the furniture, format and arch of lights display some intensive French classical artistic atmosphere. The color tone of the background wall centers on milky white and brown yellow. The wall has clear lines, providing sufficient displaying space for soft decoration. Sofa, tea table and dining chairs all attract people's attention with their peculiar European palace style and metal carving technique. The symmetrical layout also presents the space beauty of classical European style. European style has high standards for metal and crystal lights. The appropriate lights root in the creation of details and pursuit for texture. The traditional illuminating functions in this project are not weakened, while functions for beautifying environment and interior space decoration are strengthened.

三盛中央花园
Sansheng Central Garden

设计单位：福建国广一叶建筑装饰设计工程有限公司
方案审定：叶斌
设 计 师：朱文力
项目地点：福建省福州市
项目面积：170 m^2
主要材料：通体砖、实木地板、圣亚米黄大理石、壁纸、喷花茶镜、皮革软包

Design Company: Fujian Guoguangyiye Construction Decorative Design and Engineering Co., Ltd.
Project Examiner: Ye Bin
Designer: Zhu Wenli
Project Location: Fuzhou in Fujian Province
Project Area: 170 m^2
Major Materials: Brick, Solid Wood Floor, Beige Marble, Wallpaper, Tawny Glass, Leather Soft Roll

在这套样板房里，设计师以打造新古典主义风格为主，采用删繁就简的形式，充分发挥开敞空间的优势，简化了欧式风格中常用的繁复线条，让优雅、精致而富有内涵的新古典主义风格可以完全地展现出来。这个设计像是一种多元化的思考方式，将怀古的浪漫情怀与现代人对生活的需求相结合，兼容华贵典雅与时尚现代，反映出后工业时代个性化的美学观点和文化品位。

For this show flat, the designer centers on neo-classical style and applies the approach of simplifying by cutting out the superfluous. Thus the advantage of the space is displayed to the extreme and the usual complex lines of European style are simplified. Elegant and extricate neo-classical style with connotations is utterly represented. This design is like some diversified thinking mode. Nostalgic romantic sensations are connected with modern people's pursuit of life, possessing both aristocracy and modern fashion, reflecting personalized aesthetic viewpoint and cultural taste of post-industrial age.

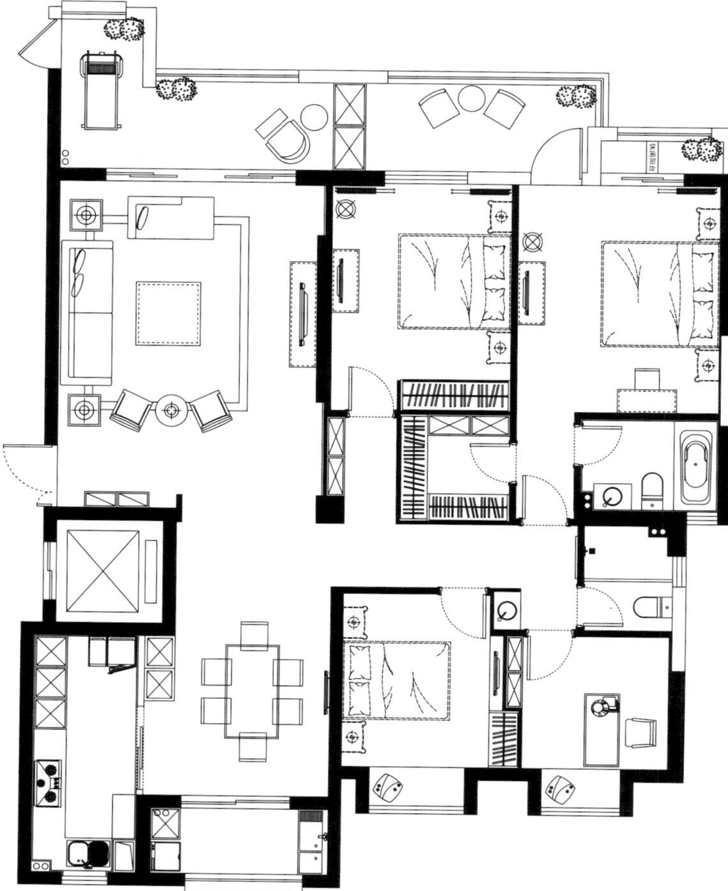

锦园新古典样板房

Jinyuan New-Classical Show Flat

设计单位：珠海空间印象建筑装饰设计有限公司
设 计 师：霍承显
软装设计：潘敏意
项目地点：广东省珠海市
项目面积：270 m²

Design Company:Zhuhai Space Impression Architectural Decorative Design Co.,Ltd
Designer: Huo Chengxian
Soft Decoration Designer: Pan Minyi
Project Location: Zhuhai in Guangdong Province
Project Area: 270 m²

本案坐落于城市的中心地带，周边有公园等城市景观，室内面积有 270 m^2，每层设置有两梯两户，室内空间南北通透，一开始就具备了奢华户型的潜质。在这套样板房中，设计师采用欧式新古典的奢华风格进行设计，以现代装修的手法重现了新古典的欧式家居意境。一眼望去，造型优雅大气的家具塑造出低调奢华的整体氛围，顶棚上的吊灯造型简洁而又有质感，背景墙上的镜面花纹呼应着欧式典雅的氛围。在整个空间中，新古典奢华风格经历了多次的推敲、完善、调整，再推敲、再完善的过程……正是这种精益求精的态度，使居室中的家居饰品可以跟整体的格调和谐统一，显现出富丽堂皇的姿态。

This project is located in the city's central area, with gardens and other urban landscape around it. The interior space is about 270m^2, with two apartments on each floor, plus one elevator, one stair. The flat with transparent interior space has the potential to become luxury flat type from the very beginning. For this show flat, the designer applies neo-classical European luxury style to reproduce neo-classical European home furnishing with modern decoration approach. Furniture of elegant and magnificent format displays low-key and luxurious whole atmosphere. The droplight from the ceiling is concise, but with texture. The mirror surface decorative pattern on the background wall corresponds with European elegant atmosphere. Inside the whole space, neo-classical luxury style goes through rounds of deliberation, perfection, adjustment and re-perfection. It is out of this constant striving for perfection, the home furnishing decorative objects are made harmonious and consistent with the whole tone, displaying magnificent posture.

公园道一号

No. One Park Road

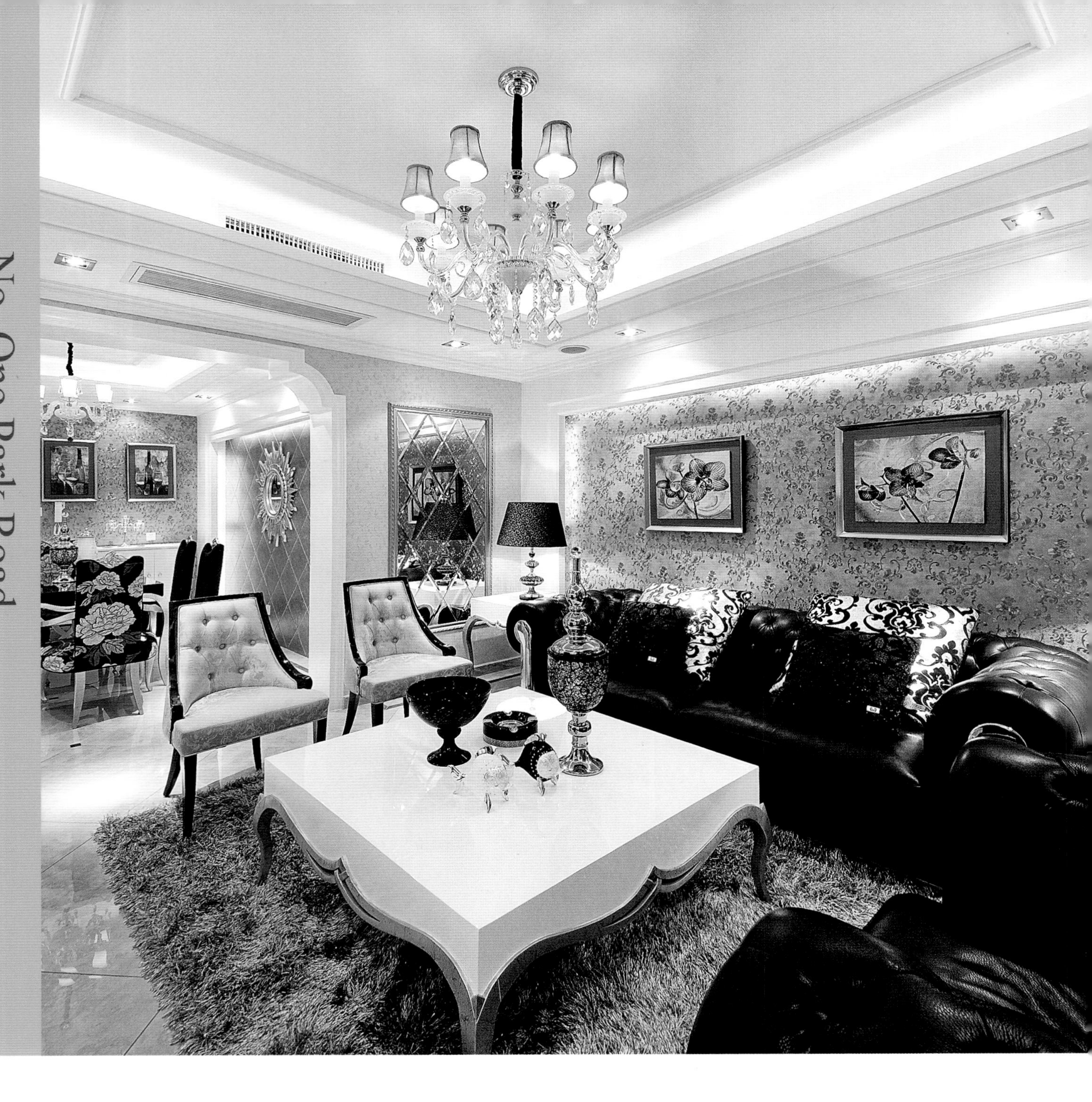

设 计 师：黄耀国
项目地点：福建省福州市
项目面积：140 m^2
主要材料：木纹大理石、仿古砖、水曲柳饰面板、明镜、壁纸、水泥漆

Designer: Huang Yaoguo
Project Location: Fuzhou in Fujian Province
Project Area: 140 m^2
Major Materials: Wood Grain Marble, Antique Brick, Ashtree Veneer, Mirror, Wallpaper, Cement Paint

本案的业主希望在享受休闲居住空间的同时，拥有现代欧式文化的内涵。因此，设计师不仅以其感性的手法诠释了欧式新古典风格的概念，而且在对空间感受的表达上有了进一步的升华。在客厅的整体色调上，设计师采用了黑与白的强烈对比来提升空间质感，又通过多种材质的相互碰撞产生丰富的视觉效果。

客厅、餐厅等公共区域的色彩和谐统一，暗紫色的窗帘上婉约的花纹展现出室内空间低调华丽之感。菱形的镜面装饰不仅在视觉上扩大了室内面积，而且与电视背景墙上的装饰图案相互呼应。卧室的设计强调温馨、舒适的睡眠环境，粉色调的床品配上背景墙上褐色的软包流露出一丝高雅的气息。儿童房整体为浅蓝色色调，采用的壁纸充满童趣，给孩子带来一个轻松自然的卧室环境。设计师善于利用不同的材料构筑出典雅而不烦琐的居住空间，同时在关键部位流露出精彩。

The property owner of this project hopes to possess the connotations of modern European culture while enjoying leisure and residential space. Thus, the designer not only interprets the concepts of European neo-classical style with perceptual approaches, but also attains some further sublimation in the expression of space perceptions. For the whole color tone of the living room, the designer applies the strong contrast of black and white to improve the texture of the space and produces abundant visual effects with the crash of various materials.

The public area such as living room and dining restaurant has harmonious color tone. The graceful pattern on the dark purple curtain displays the low-key and magnificent feel of the interior space. Diamond mirror-surface decoration not only enlarges the interior space visually, but also corresponds with the decorative patterns on the TV's background wall. The bedroom design emphasizes warm and cozy sleeping environment. The combination of pink bedding accessories and brown soft roll on the background wall display some elegant atmosphere. The children's room has light blue color tone with wallpaper of child interest creating some relaxing and natural bedroom environment. The designer is good at making use of different materials to construct some elegant but not cumbersome residential space, with some splendid elements in the key parts.

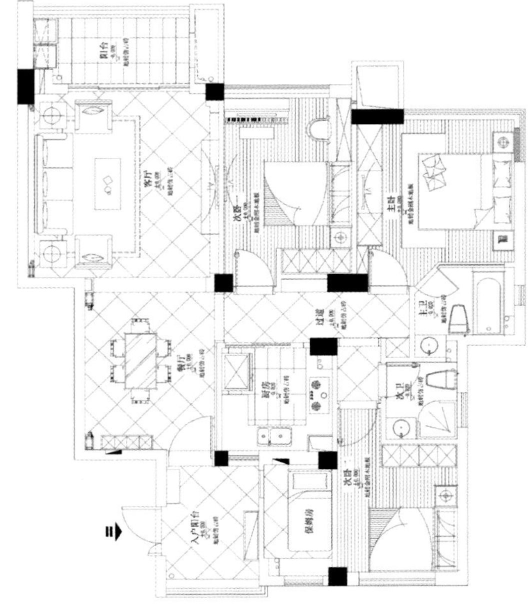

长春益田枫露丹堤样板房

Changchun Yitian Fenglu Dandi Show Flat

设计单位：深圳市艺鼎装饰设计有限公司
设 计 师：王锟
项目地点：吉林省长春市
项目面积：120 m^2

Design Company: Shenzhen Yiding Decoration and Design Co., Ltd.
Designer: Wang Kun
Project Location: Changchun in Jilin Province
Project Area: 120 m^2

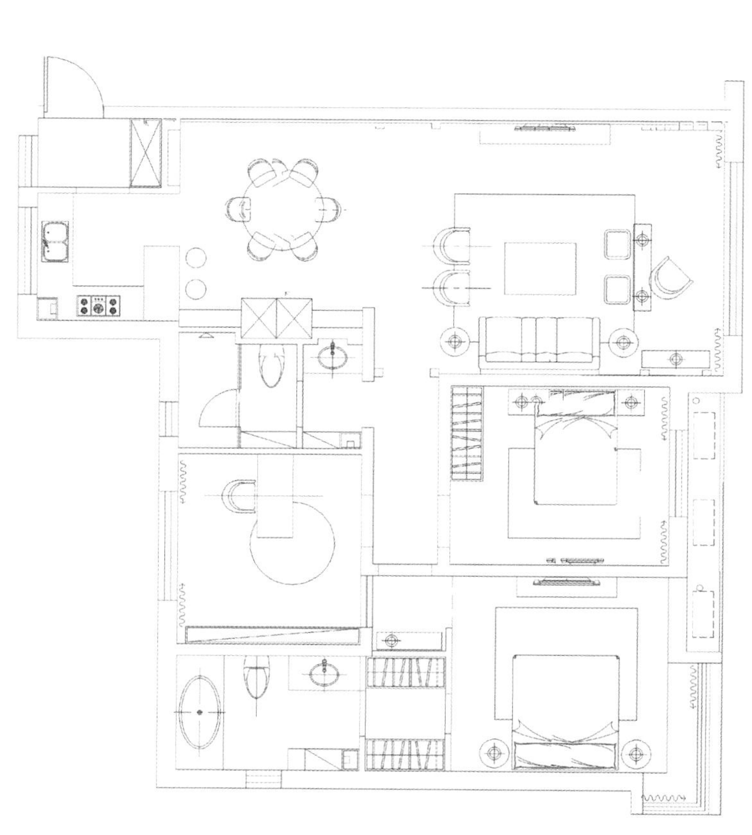

本案选用白色系为主的色调来装饰空间，奠定了空间优雅、柔和的格调。客厅中大面积的白色实木线条造型的背景墙装饰突显出欧式新古典的情怀，而悬挂在其中央的挂画以黑色为底色，强调了空间带给人的视觉冲击力。金色的画框将两种颜色和谐地融合在一起，并透出一缕古典优雅的气息。正对着挂画的墙面上，设计师采用了材质反差极大的印花镜面来装饰，简洁大方的花纹作为墙面的主体，又为空间带来了一丝浪漫的气息，配合着新古典风格的家居，呈现出地道的欧式风情家居。

餐厅的设计延续了客厅的风格特征，色调以黑、白的对比为主，在简洁的立面设计基调下，挑选了同样简约、时尚的家具，让整体空间看起来干净整洁，又温馨时尚。卧室空间饰以淡雅的壁纸，让人置身于一个柔和的空间，舒适的床品以及典雅的家具一同展现出纯净的色调感，营造出适合睡眠的质感空间。

This project applies white color tone to decorate the space, establishing graceful and soft tones for the space. Inside the living room, the background wall with large area of white solid wood lining highlights neo-classical European emotional appeals. The painting hung in the middle has black as the ground color, emphasizing on the visual impact that the space leaves people with. Gold painting frame combines these two colors harmoniously, revealing some classical and elegant atmosphere. On the opposite wall, as contrast, the designer applies printing mirror surface as the decoration. Concise and magnificent decorative pattern becomes the body part of the wall, while bringing some romantic atmosphere for the space. Accompanied with furniture of neo-classical style, the design displays authentic home furnishing of European emotional appeals.

The design of the dining hall continues the style and characteristics of the living room, with black and white as the color tone. Based on concise elevation design, the designer selects concise and fashionable furniture to make the whole space appear tidy, clean and warm. The bedroom is decorated with elegant wallpaper. It is like that people are being inside a soft space. Comfortable bedding accessories and elegant furniture together display some pure color tone, creating some texture space quite appropriate for sleeping inside.

都市印象

Urban Impression

设计单位：威利斯设计有限公司
设 计 师：蒋娟
项目地点：江苏省苏州市常熟衡泰花园洋房
项目面积：130 m^2
主要材料：皮纹砖、仿木纹砖、爵士白大理石、车边镜、壁纸、地板

Design Company: Willis Design Co., Ltd.
Designer: Jiang Juan
Project Location: Suzhou in Jiangsu Province
Project Area: 130 m^2
Major Materials: Leather Grain Brick, Imitation Wood Grain Brick, Jazz White Marble, Decoration Mirror, Wallpaper, Floor

本案为三室两厅的平层公寓，户型动静分区合理，设计师对其进行了微小的改造，使其功能更为完备。进门处增加了小小的门厅空间，保证了业主生活的私密性。厨房与卫生间之间也进行了简单的改造，增加了冰箱的位置，使布局更为合理。

会客区吊顶部分打破了传统的横平竖直，根据不同功能区的划分进行设计，石膏线与原木色饰面板相结合，形成一个不规则的多边形，视觉上丰富多变、别具一格。进门右手处是餐厅，一侧利用墙壁设计成简单的装饰酒柜，餐桌、座椅、色彩斑斓的装饰画、吊灯构成餐厅的全部，简约而温馨。

客厅非常注重层次感，尤其是电视背景墙部分，采用明镜、原木色饰面板、无框玻璃门、咖啡色皮纹砖、百叶窗等不同的材质巧妙组合而成，与休闲区隔而不断，又延展了视觉空间。

卧室的环境温馨安逸，主卧没有使用大灯，而是通过筒灯、壁灯、隐藏灯等来营造舒适的睡眠环境。墙壁的装饰也仅采用装饰画，素雅宁静。

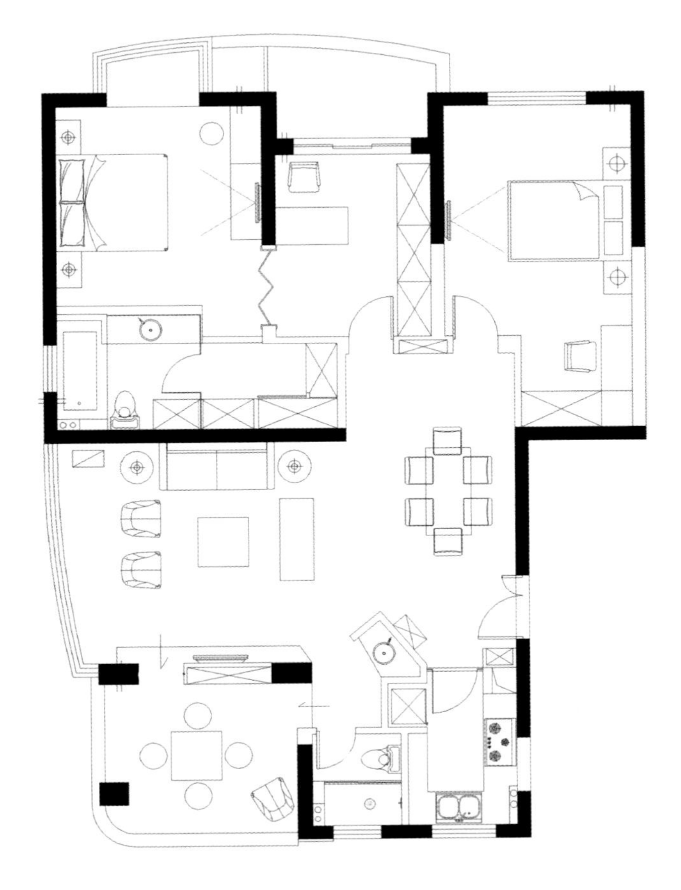

This is an apartment with three bedrooms and two halls, with proper distinction between the active space and the quiet space. The designer made some minor transformation for the space to make the functions much more complete. There is a small hallway at the entrance, ensuring the privacy of the property owner's life. The designer also makes some simple transformation for the space between the kitchen and the washroom, setting a refrigerator here, with more appropriate layout.

The ceiling of the living room breaks through the traditional horizontal or vertical design, but carries out design with division of different functional areas. Gypsum lining is connected with the log wood color veneer, producing some irregular polygon, with visual variations and quite peculiar. On the right side of the entrance is the dining hall. There is a simple decorative wine cupboard here. Dining table, dining chairs, colorful decorative painting and drop lights compose the whole of the dining hall, simple but full of warmth.

The living room focuses on the layers, especially the TV background wall, which applies the ingenious combination of different materials such as bright mirror, veneer of log wood color, frameless glass door, leather grain brick of coffee color, window shutter, etc. The TV background wall is separated with the leisure zone while connects with each other, visually expanding the space.

The environment of the bedroom is warm and comfortable. The master bedroom does not use grand lights, but apply the combination of tube lights, wall lamps and concealed lights to create the cozy sleeping environment. The wall only applies decorative painting, which appears tranquil and elegant.

万科金域蓝湾 缤纷印象样板房

Vanke Jinyu Blue Bay Colorful Impression Show Flat

设计单位：广州共生形态工程设计有限公司
WWW.COCOPRO.CN
项目地点：广东省佛山市南海区

Design Company: Guangzhou C&C Design Co., Ltd.
WWW.COCOPRO.CN
Project Location: Nanhai District, Foshan in Guangdong Province

本案是针对开发商的标准装修样板房，设计师通过软装中的色彩和材质的搭配来表现不同的气质与主题。设计师采用欧式新古典风格对空间进行诠释，利用不同材料产生的不同心理感受来优化空间，为居室带来典雅、高贵的质感。一步入客厅，一股优雅、清新的气息便扑面而来，纹样简洁的地毯、璀璨的水晶灯和典雅的家具传承了新古典风格的文化底蕴和美感，使古典和现代完美地融合。

欧式的居室在设计上更加注重对浪漫氛围的营造。主卧空间选用了缎面壁纸装饰墙面，丝绸般的质地带给人舒适、华贵之感，而在色彩上与床品对比鲜明，丰富、跳跃的颜色活跃了空间的气氛。设计师注重对细节的刻画，抱枕、灯具、窗帘的搭配相得益彰，无一不展示着业主对美好生活的追求。次卧采用冷色调营造出素雅、淡然的空间氛围，衣橱的线条为简洁的曲线，符合现代的审美观，优雅又大气，为空间带来异域的风情。设计师通过缤纷印象的设计主题赋予标准化住宅一种全新的色彩感官，让生活不再单调，而是充满精彩非凡的体验。

This project aims at decorating show flat according to the standards of the project developer. The designer tries to display different temperament and themes according to the combination of colors and materials in the soft decoration. The designer carries out interpretation for the space with neo-classical European style and optimizes the space with different psychological impressions created by different materials, creating some elegant and noble texture for the residential space. Once you step into the living room, you can get completely impressed by the elegant and fresh atmosphere. Carpet of concise pattern, dazzling crystal lights and elegant furniture inherits neo-classical style cultural connotations and aesthetic beauty, integrating classical style and modernity perfectly.

European residence emphasizes more on the creation of romantic atmosphere. The master bedroom selects satin wallpaper to decorate the wall. The silk texture would impress people with its comfort and magnificence, while forming sharp contrast with bedding accessories. Abundant and bouncing colors activate the atmosphere of the space. The designer focuses on the description of details. Sofa pillows, lights and curtains bring out the best in each other, all displaying the property owners' pursuit for happy life. The subordinate bedroom applies cold colors to create elegant and light space atmosphere. The cupboard has concise lining, in accordance with modern aesthetic standard, elegant and magnificent, creating some exotic temperament for the space. Through design themes of colorful impressions, the designer entrusts the standard residence with some brand-new color sense organs, making life not that monotonous, but with brilliant and uncommon experiences.

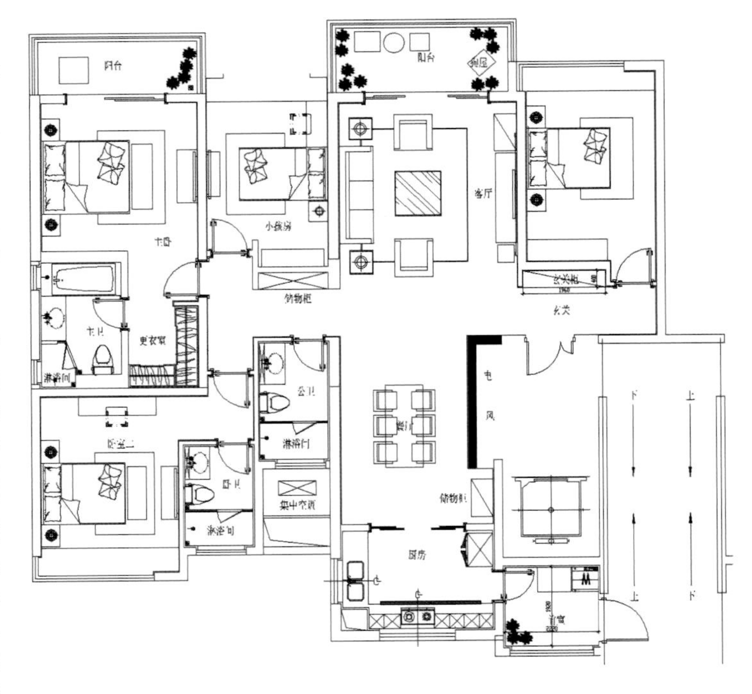

再续奢华

To Continue the Luxury

设 计 师：龚德成
项目地点：深圳
项目面积：160 m^2
主要材料：阿波罗大理石、特级大花白大理石、玻化砖 、进口壁纸、软包、水晶灯、手工地毯、木雕花、茶镜磨花

Designer: Gong Decheng
Project Location: Shenzhen
Project Area: 160 m^2
Major Materials: Apollo Marble, Special White Marble, Vitrified Brick, Imported Wallpaper, Soft Roll, Crystal Light, Handmade Carpet, Wooden Carving, Polished Tawny Mirror

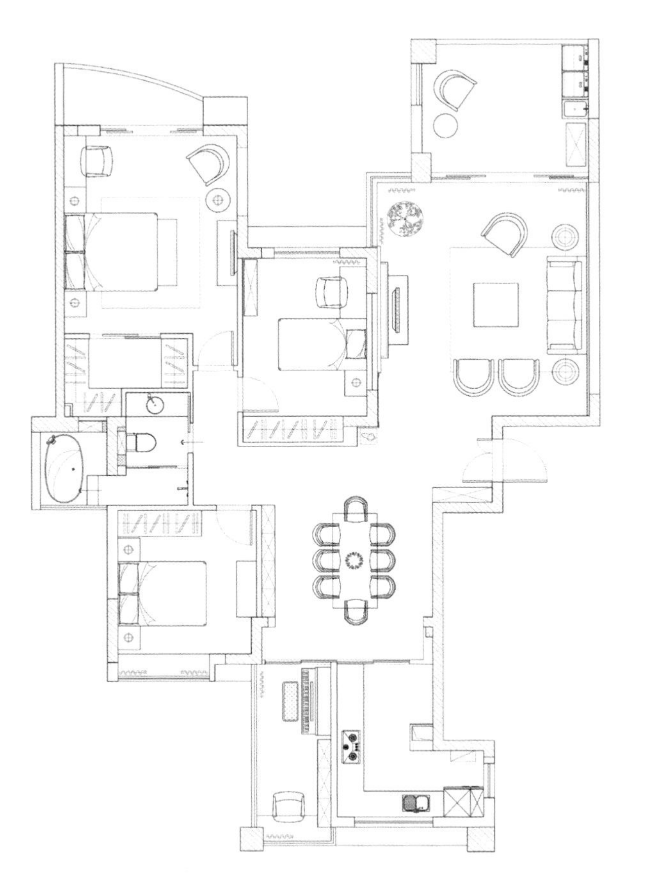

水晶吊灯、烛台、镶银边的餐具、浅灰色的壁纸、镶银边的欧式新古典风格家具以及手工绣花羊毛地毯，充分强调了空间色彩中深色与浅色的对比、冷色与暖色的搭配，让业主在整体环境中体会到了淡淡的华丽感。设计师试图打造一种低调的生活方式，这也是业主所追求的一种高品质的生活方式，空间的设计展现出一种低调的奢华之美。

Crystal droplights, candlestick, table ware of silver lining, light gray wallpaper, Neo-Classical European furniture of silver lining and hand-made embroidery wool carpet all emphasize the contrast of bright and dark colors inside the space, and the collocation of cold and warm colors. Thus the property owner can get some slight magnificent sensations in the whole environment. The designer tries to create some low-key life style, which is also the high-quality life style that the property owner aspires for. The design of the space displays some low-key luxurious aesthetic feel.

CLUB

南阳样板间

Nanyang Show Flat

设计单位：嘉道设计 • 老鬼工作室
主设计师：老鬼
参与设计：杨凯、吴坚
项目地点：河南省南阳市
项目面积：400 m^2
投 资 商：河南宛西制药股份有限公司
开 发 商：南阳财富置业有限公司
主要材料：涂料、陶瓷锦砖、文化石、瓷砖

Design Company: Jiadao Design • Lao Gui Studio
Leading Designer: Lao Gui
Associate Designers: Yang Kai, Wu Jian
Project Location: Nanyang in Henan Province
Project Area: 400 m^2
Investor: Henan Wanxi Medicines Co., Ltd.
Developer: Nanyang Fortune Home Buyers Co., Ltd.
Major Materials: Painting, Ceramic Mosaic Tile, Cultural Stone, Ceramic Brick

河南嘉道装饰公司

在这个喧嚣的社会中，人们在生活上不断地追求典雅、尊贵的生活品质。本案采用西班牙式的温情摇滚乐的设计手法，对西班牙的地域文化、建筑元素等进行了提炼。将西班牙的特色文化马蹄门元素引入室内空间，体现了现代温情的生活气息。地面的拼花体现了斗牛士的精神，用棱角分明的铺法，将室内摇滚乐的气氛表现得淋漓尽致。

在功能设计上，除了原建筑的结构、特征以外，在空间的使用性上做了最大化的优化，勾画出了神秘、内敛、沉稳、厚重、色彩古朴又不失贵族气质的居室氛围，让业主深切地体会到空间的温情和舒适，让家成为真正的生活舞台。

在软装搭配上，选用镶嵌有金属的家具，金银质感饰面的饰品起到了画龙点睛的作用，不但体现了尊贵的生活品质，也将西班牙式的摇滚乐文化带入了家庭生活之中。

在美妙的西班牙式温情摇滚乐中，人们可以感受到阳光、沙滩的浪漫生活，为了体现这段浪漫的“旅程”，在户外花园的设计上，设计师采用了赤陶、粗糙的毛

石、黑色的铁艺雕花等元素，充分体现出了地中海的阳光海岸情调，让这个豪宅更加温情。业主可以到观景台上，坐在蓝白相间的太阳伞下，喝着百年红酒，体验着休闲、自然的浪漫温情生活情调，这是设计师在作品中的得意之处，也展现了业主的生活追求与向往。为了增加生活的乐趣与情调，设计师还在户外空中花园中设置了儿童娱乐区，营造了人与人之间

的情感交流的空间。

用设计师自己的一句设计名言来概括这个作品的精髓，即：“作为一个设计师要学会挖掘生活细节，挖掘生活的情感和经历，同时还要学会反省生活中的过错，这样在设计一个新作品的时候，才不会把自己的情绪强压给他人，才能真正体现作品的空灵之美。”

people have always been pursuing for elegant and noble life quality. This project applies the design approach of Spanish warm rock music to extract something from Spanish regional culture and architectural elements. In the interior space, there is this horseshoe style door with specific Spanish cultural elements, representing the modern warm life atmosphere. The pattern on the floor displays the spirit of matador. The pavement of clear edges and corners display the atmosphere of interior rock music to the utmost.

As for the functional design, apart from the construction and features of the original architecture, the designer optimizes the usability of the space, depicting some mysterious, sedate, archaic but aristocratic interior atmosphere. The property owner can truly observe the warmth and comfort of the space, making home a real life stage.

For the collocation of the soft decoration. The designer selects furniture with metal. The ornaments of gold and silver texture veneer make the

finishing point, which not only displays the noble life quality, but also brings Spanish rock culture into family life.

In the beautiful Spanish warm rock music, people can sense the romantic life with sunshine and beach. For the presentation of this romantic journey, for the design of outdoor garden, the designer applies terra-cotta, coarse rubble, black iron carving and other elements to full display the sunny coastal charms of Mediterranean, making the mansion appear even warmer. The property owner can come to the observation platform, sitting under the blue and white umbrella, drinking red wine of 100 years and experiencing casual, natural and romantic warm life tones. And that is the part the designer is quite proud of, while displaying the property owner's life pursuit and longings. In order to add some more life interests and emotional appeals, the designer set a children's entertainment area in the outdoor air garden, creating some space for communications among people.

We can summarize the core essence of this work with the famous remark by the designer himself: "As a designer, you need to learn to dig out the

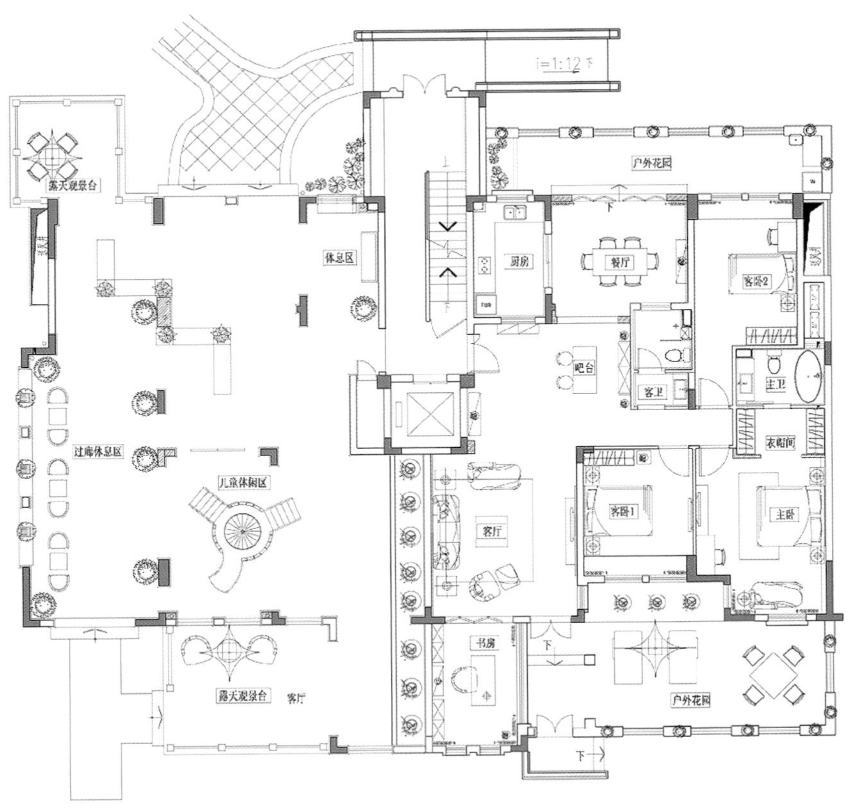

life details, get out the emotions and experiences in life and while at the same time try to reflect on the wrong-doings in life. Only upon doing this in designing a new work, one would not impose his own emotions upon other people can get to display the intangible beauty of the work."

相融 Integration

设计单位：福建国广一叶建筑装饰设计工程有限公司

方案审定：叶斌

设 计 师：谢颖雄

项目地点：福建省福清融侨城

项目面积：130 m^2

主要材料：大理石、软包、砂岩

Design Company: Fujian Guoguangyiye Construction Decorative Design and Engineering Co., Ltd.

Project Examiner: Ye Bin

Designer: Xie Yingxiong

Project Location: Rongqiaocheng Property in Fuqing, Fujian Province

Project Area: 130 m^2

Major Materials: Marble, Soft Roll, Sandstone

对这套住宅进行设计时，设计师抓住业主喜欢时尚、追求高品质生活的特点，在整体布局和材料的运用上，巧妙地把欧式古典元素以现代的形式表现出来。

雅致的背景造型、考究的配饰、精湛的工艺，均是现代与古典相融的结晶，让业主的审美情趣在优美的线条、和谐的色调中自然流露。从入口到客厅再到餐厅，空间流畅通明，名贵内敛的浅色调天然石料是地面和墙身装饰的主要材料，它们与水曲柳刷白造型墙壁形成鲜明而自然的对比。纯净的米色造型框架奠定了居室的基调，设计师在装饰造型中放弃传统欧式风格中繁复的曲线与雕花，采用明快的直线条勾勒出了空间的清爽气质。

卧室与客厅相比较，个性更鲜明。主卧选择了较为沉稳的色系，家具也是大角度宽松式的。旁边另一扇门内的女孩房色彩温馨浪漫，门厅旁的男孩房在配饰上则以条纹为主，体现出一股阳刚之气。欧式元素经过设计师的严格筛选，被恰到好处地应用于各个空间的装饰中，使整所房子的格调统一起来。

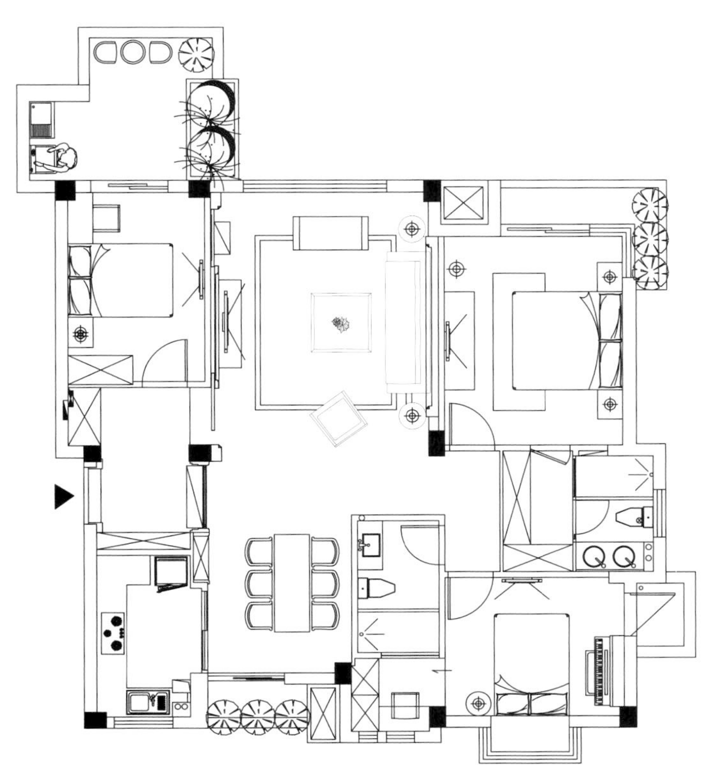

While carrying out design for this residence, the designer captures the property owner's characteristics of fancying fashion and pursuing high-quality life and ingeniously displays classical European elements with modern forms in the whole layout and the application of materials.

Elegant background format, exquisite decorative ornaments and consummate techniques are all the product of integration of modern and classical design which displays the aesthetic temperament and interest of the property owner in elegant lines and harmonious color tones. From the entrance to the living room, till the dining hall, the space is fluent and bright. Natural stone of light color tone with aristocratic temperament is the major material for the ground and wall decorations, which produces some distinct and natural contrast with the wall of ash-tree white format. Pure beige color format frame sets the tone of the space. For the decorative format, the designer forsakes complicated curves and carvings in the traditional European style, depicting the clear temperament with vibrant straight lines.

Compared with the living room, the bedroom selects much more sedate colors. And the furniture appears relaxing and comforting. The girl's room on the side has warm and romantic colors. The boy's room beside the hall focuses on stripes on the decorative ornaments, expressing some masculine temperament. Through serious selection of the designer, the European elements are appropriately applied in the decoration of various spaces, which unifies the whole residence.

江南中央美地

Jiangnan Zhongyang Meidi

设计单位：无界空间陈设机构
软装设计：吴华
项目地点：福建省南平市
项目面积：95 m²
摄 影 师：施凯

Design Company: Wujie Space Layout Institution
Soft Decoration Designer: Wu Hua
Project Location: Nanping in Fujian Province
Project Area: 95 m²
Photographer: Shi Kai

本案遵循新古典风格，主要元素有华贵的颜色、镜面和不锈钢的材质饰面。设计师摒弃了过于复杂的肌理和装饰，简化了线条，通过结合现代人的审美观体现了新古典主义不一样的优雅格调。样板间中银色的运用彰显奢华的格调，冰冷的高贵也体现了唯美的新奢华风。

宽亮温馨的客厅，别致的玻璃吊灯，锃亮的镜面，简单几笔便勾勒出空间浓浓的小资情调。咖色、黑色、银色为主色调，与白色糅合，使色彩看起来明亮、大方，使整个空间给人以开放、宽容的非凡气度，让人丝毫不感局促。本案将怀古的浪漫情怀与现代人对生活的需求相结合，兼容华贵典雅与时尚现代，反映出后工业时代个性化的美学观点和文化品位。

主卧室弥漫着女主人精致高雅的味道，空间整洁而不乏温馨的生活气息。新古典主义传承了古典主义的文化底蕴、历史美感及艺术气息，兼容华贵典雅与时尚现代，将繁复的家居装饰凝练得更为简洁精雅，为硬而直的线条配上温婉雅致的软性装饰，将古典美注入简洁实用的现代设计中，使家居装饰更有灵性，让古典的美丽穿透岁月，在我们的身边活色生香。

设计师突破了循规蹈矩的家居布置，让空间充满了活力和无拘束的气氛。空间饰品的摆放干净利落，与不规则的家具布置相映成趣。设计师很好地利用了空间的局限性，又使整体效果别具一格。

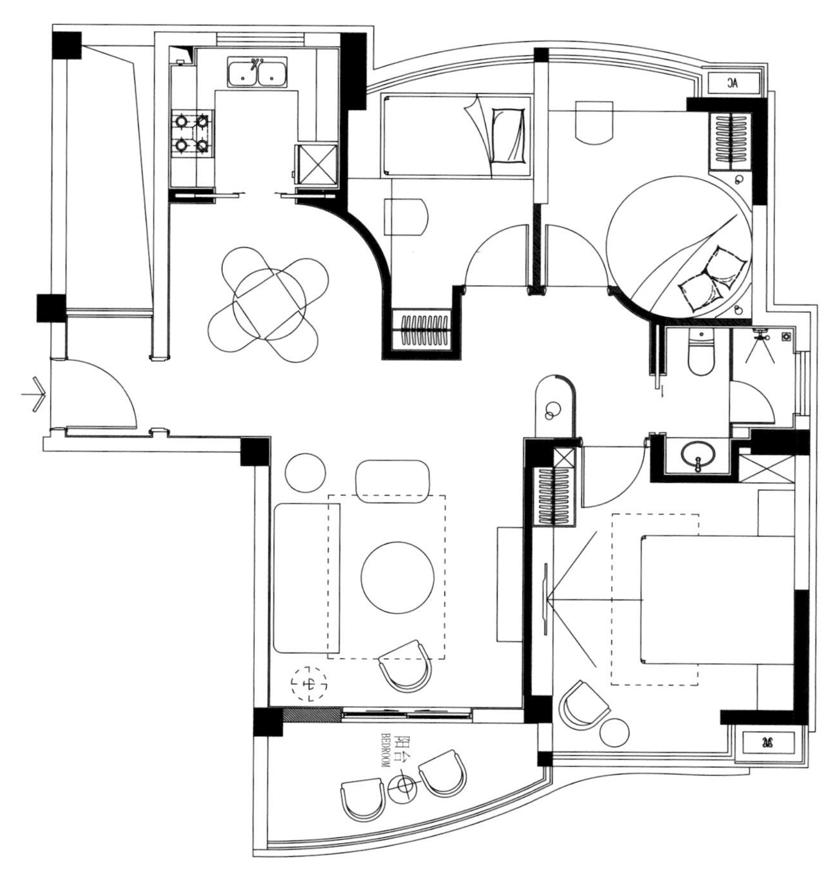

This project follows neo-classical style, with main elements of noble colors, mirror surface and stainless steel veneer. The designer forsakes much too complicated texture and decorations, simplifies lines and displays the elegant tone of neo-classical style through the combination with aesthetic standard of modern people. The application of silver color inside the show flat displays luxurious tone. It also displays aesthetic new luxury style.

Broad and warm living room, peculiar glass droplights, bright mirror surface... Several brushes would produce the intensive petty bourgeoisie sensations inside the space. The color tone centers on coffee, black and silver, combined with white. The colors appear bright and generous, creating some open and generous temperament for the whole space, which does not make people feel narrow at all. This project combines nostalgic romantic sensations and modern people's requirements for life, combining fashion, elegance and modernity and reflecting individual aesthetic viewpoints and cultural tastes of post-industrial age.

The master bedroom is pervasive with the exquisite and elegant taste of the hostess. The space is clear and tidy but with no less warm life atmosphere. Neo-classicism inherits the cultural connotations, historical aesthetic beauty and artistic atmosphere of classicism. With aristocratic elegance and fashionable modernity, the design makes complicated home furnishing appear concise and elegant, accommodates hard and straight lines with elegant soft

decorations and instills classical beauty into concise and practical modern design, thus making the home furnishing dynamic and allowing classical beauty to penetrate through life, bringing out bright colors around us.

The designer breaks through the regular home furnishing layout, creating lively and unbridled atmosphere inside the space. The layout of the ornaments is neat and tidy, contrasting finely with the irregular furniture. The designer makes well use of the limitation of the space, thus making the whole space quite spectacular.

白色新古典

White Neo-Classical Design

设计单位：武汉郑一鸣室内建筑设计

设 计 师：郑一鸣、吴锦文

项目地点：湖北省武汉市

项目面积：260 m^2

主要材料：霸王花大理石、雅士白大理石、壁炉、银箔陶瓷锦砖、艺术壁纸、银镜

Design Company: Wuhan Zheng Yiming Interior Architectural Design Firm

Designers: Zheng Yiming, Wu Jinwen

Project Location: Wuhan in Hubei Province

Project Area: 260 m^2

Major Materials: Jazz White Marble, Fireplace, Silver Foil Mosaic Tile, Artistic Wallpaper, Silver Mirror

本案为湖景房，周边视野开阔、景色优美，设计师选用新古典风格来展现其恬静与时尚。新古典主义风格，是在古典主义风格的基础上做减法，减去历史的沉重，削掉一层油彩，抹去金银的俗媚，如此简单的奢华，让人沉醉。

居室的整体色调为米白色，搭配上白色的新古典风格家具显得清新、淡雅。娇媚的花朵为空间增添了一抹春天的气息，也使室内气息更显柔美。规则的方形地砖边界明显，更加延伸了空间的尺度感，凸显出空间落落大方的气质特点。同时，室内空间再加上一点现代气息的几何图案的点缀，以及一些现代人的个性，将新古典风格的美展现得淋漓尽致，凸显出简单中透着清新的设计风格。本案用其优雅的气质征服了时下追求内涵和品质生活的精英一族，让他们的心灵在此获得回归感。

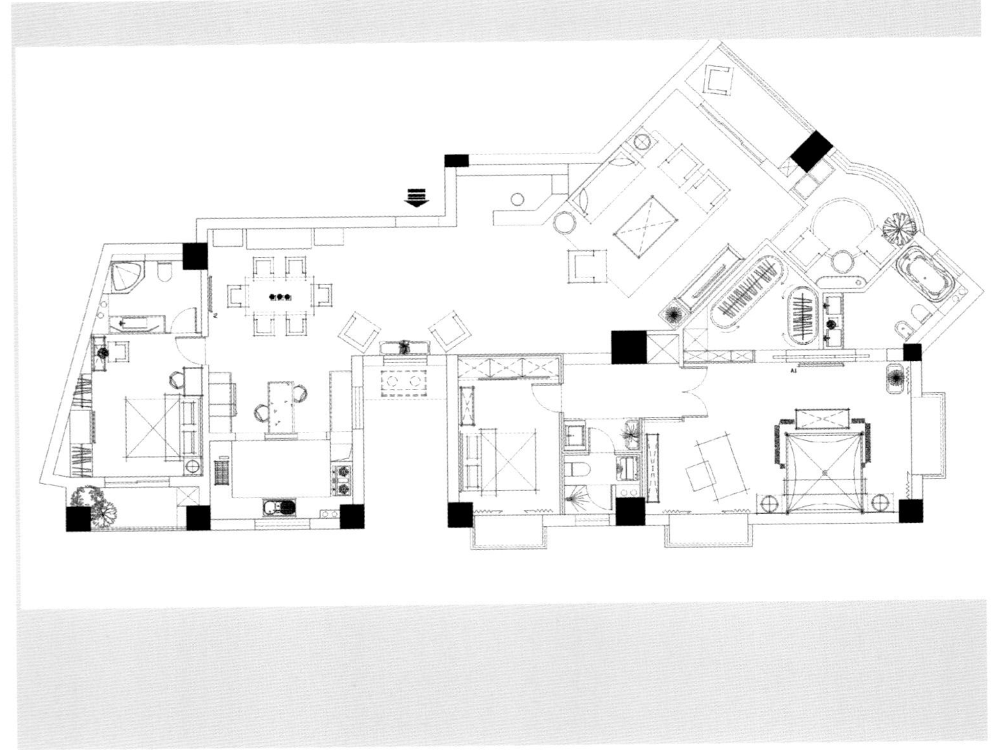

This is a project with lake views, with broad views around and nice landscape. The designer selects neo-classical style to display the calmness and fashion. Neo-classical style is to make subtractions based on classical style, subtracting the heaviness of history, erasing some paint and the vulgarity of gold and silver. Such simple luxury is so enchanting.

The whole color tone of the residence is beige white, appearing fresh and elegant accompanied with white neo-classical style furniture. The charming flowers create some spring atmosphere for the space, which makes the interior atmosphere softer and nicer. The regular square tile has clear boundaries, further expanding the scale of the space, highlighting the generous temperament of the space. At the same time, the interior space is added with some geometric pattern ornaments of modern atmosphere and some modern characteristics, displaying the beauty of neo-classical style to extremes, setting off the fresh design style within this simplicity. With its elegant temperament, this project wins the heart of the elite who aspire for connotations and quality life, giving their heart some sense of returning.

南京彩云居

Nanjing Iridescent Clouds Mansion

设计单位：大匀国际空间设计
设 计 师：林宪政
软装设计：太舍馆贸易有限公司
项目地点：江苏省南京市
项目面积：110 m^2
开 发 商：中粮地产南京有限公司
主要材料：古典米黄大理石、枫桦木地板、壁纸

Design Company: Symmetry Space Design
Designers: Lin Xianzheng
Soft Decoration Designer: MOGA DECO Co., Ltd.
Project Location: Nanjing in Jiangsu Province
Project Area: 110 m^2
Developer: Cofco Property (Nanjing) Co., Ltd.
Major Materials: Classical Beige Marble, Maple and Birch Wood Floor, Wallpaper

净心莲之境

莲花，代表着一种至纯至清的理想境界，它以“出淤泥而不染，濯清涟而不妖”的品性受到人们的喜爱，而其清美的外在形态以及淡淡的芳香都让人一见倾心。本案以莲为设计主题，以白色为设计主调，在空间中充分展现了一种清雅、高贵之气。

客厅的布局简洁、大方，色调以灰色系配合白色为主，同色系的布艺沙发和摆件相映成趣，带给人朴素、淡雅的空间体验。墙上的挂画是经过精心挑选的，含蓄的笔锋让莲花自然清新的形象深入人心，更为客厅营造出温馨之感。餐厅展现出业主享受生活的精致格调，白色餐桌上玻璃质感的烛台和顶棚上华丽的水晶灯相互呼应，给空间带来一片璀璨之色，细致光洁的餐具盛放着美味的食物，让人颇有食欲。书房与过道的分隔造型墙以及过道的对景搭配得相得益彰。主卧的绿丝绒床背靠为空间增添了一抹亮色，也更加映衬了主题白莲风格的素雅。

Pure Lotus World

Lotus represents some pure and clear ideal world. It wins people's love with its characteristics of “maintaining pure out of the sludge and noble after being washed in the ripples”. Its pure posture and light fragrance win people's love at first sight. This project has lotus as the design theme and white color as the homophony, presenting some elegant and noble temperament in the space.

The layout of the living room is concise and liberal. The color tone centers on grey and white. Same color cloth sofa and accessories form delightful contrast, creating some simple and elegant space experiences. The pictures on the wall are after careful selection. Implicit tip of a brush make the lotus with natural and fresh image enjoy popular support, while adding some warm impressions for the space. The dining hall displays the property owner's exquisite style of enjoying life. The candlestick with glass texture on the white dining table echoes with the magnificent chandeliers on the ceiling, creating some resplendent color for the space. There is delicious food in the clean and bright tableware, arousing people's appetite. The separation format wall in the study and the corridor and the opposite scenery in the corridor bring out the best in each other. Green velvet bed back in master bedroom echoes interestingly with the simple but elegant lotus decoration.

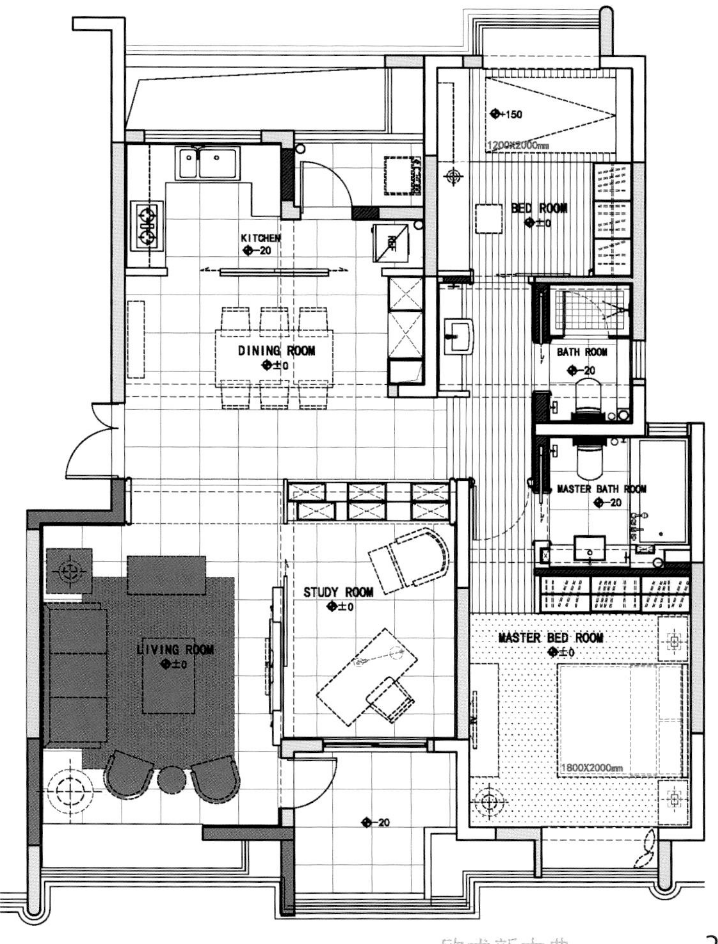
KITCHEN
DINING ROOM
BED ROOM
1200X2000mm
BATH ROOM
MASTER BATH ROOM
STUDY ROOM
LIVING ROOM
MASTER BED ROOM
1800X2000mm

尚湖中央花园

Shanghu Lake Central Garden

设计单位：威利斯设计有限公司
设 计 师：李琮
项目地点：江苏省常熟市
项目面积：140 m²
主要材料：爵士白大理石、石膏板、软包、硬包、陶瓷锦砖、银镜、地面拼花、壁纸、浅色地板

Design Company: Willis Design Co., Ltd.
Designer: Li Cong
Project Location: Changshu in Jiangsu Province
Project Area: 140 m²
Major Materials: Jazz White Marble, Plasterboard, Soft Roll, Hard Roll, Ceramic Mosaic Tile, Silver Mirror, Parquet, Wallpaper, Light Color Floor

本案为三室两厅的平层公寓，设计师以白色为基调，用欧式新古典风格的设计手法来营造现代奢华的感觉。进门处采用整体定制柜大面积装饰墙面，增加其储物功能。挂壁式鱼缸成为进门一景，在风水上蕴涵了一定的寓意，同时视觉上丰富了墙面的层次。客厅与餐厅在一条主动线上，使视觉空间更为开阔，在墙面造型的设计上两者均采用了银镜与壁纸做成的对称造型来装饰。中间以简单的吧台造型作为过渡，供业主休闲茗茶。由于层高及梁位置的局限，顶棚采用镜面铺设，拉伸空间，消除压抑感。

客厅电视背景墙由咖啡色硬包与银色金属感陶瓷锦砖构成，配合爵士白大理石以欧式线条抄边，同时，设计师巧妙地将次卧房门隐藏其中，与电视墙融为一体，增加了神秘感。新古典风格的咖啡色皮质沙发与电视墙硬包色彩相互辉映，更显空间的大气和家居生活的奢华。

卧室铺设实木地板，床背景墙采用软包，其他采用素雅的壁纸装饰，搭配白色欧式梳妆台，实用、简约又不失高雅。卫浴间虽然空间较小，但经过设计师别出心裁的设计，功能完善，原有墙面被改为透明玻璃后也使整体卧室空间更为宽敞明亮。

This is a leveling apartment with three bedrooms and two living rooms, with white as the color tone. The designer applies neo-classical European design approach to create modern and luxurious sensations. The entrance area applies whole custom-made cabinet to decorate large area of wall to add some storage functions. The wall mounted fish tank becomes scenery upon entering the door, with some connotations in the geomancy aspect, while visually enriching the layers of the wall. The living room and the dining hall are on the same moving line, visually broadening the space. As for the aspect of wall format, both rooms apply the symmetrical format made of silver mirror and wallpaper as the decoration. In the middle, there is a simple bar counter applied as the transition, for the property owner to spend some leisure time and enjoy some tea. Due to the limitation of level height and location of the girder, the ceiling applies mirror surface to stretch the space and eliminate senses of oppression.

The TV background wall inside the living room is composed of coffee hard roll and silver ceramic mosaic tile of metal texture, accompanied with jazz white marble and European lining. While at the same time, the designer ingeniously sets the secondary bedroom inside which integrates with the TV background wall, adding some mysterious atmosphere. Neo-classical coffee leather sofa and the TV background wall's hard roll bring out the best in each other, further bringing out the magnificence of the space and the luxury of residential life.

The bedroom ground is paved with wood flooring. The bed's background wall applies soft roll, while other areas apply simple but elegant wallpaper as the decoration, accompanied with white European dresser, practical, concise but graceful at the same time. The washroom is small, but after the ingenious design of the designer, it has consummate functions. After the original wall is changed into transparent glass, the whole bedroom is made open and bright.

HOUSE 1992 太和广场

HOUSE 1992 Taihe Square

设计单位：杭州麦丰装饰设计有限公司
设 计 师：陆宏
项目面积：170 m²
主要材料：仿古砖、饰面板、壁纸、实木线条

Design Company: Hangzhou My Home Design Co., Ltd.
Designer: Lu Hong
Project Area: 170 m²
Major Materials: Antique Brick, Veneer, Wallpaper, Solid Wood Lining

CCTV

本案是四室两厅的空间布局，整体色调给人明朗开阔之感。设计师采用新古典风格进行装饰，遵循以简饰繁，从整体到局部的细致刻画。室内设计保留了新古典风格对材质、色彩的运用，从空间中仍然可以很强烈地感受到传统的历史痕迹与浑厚的文化底蕴，同时又摒弃了过于复杂的肌理和装饰，简化了线条。在这里，无论是家具还是配饰均以其优雅、唯美的姿态，平和而富有内涵的气韵，描绘出业主高雅的贵族身份。

居室整体以浅色调为主，使空间看起来明亮、大方，给人以开放、宽容的非凡气度，让人丝毫不觉局促。深木色的家具配上白色和浅木色等颜色的配饰，不再让人觉得沉闷，而在视觉上让人顿生焕然一新的感觉。家具以古朴的色调、天然的材质和深沉而含蓄的风格体现出崇尚自然的作风，家具的造型、纹路、雕饰和色调细腻高贵，耐人寻味之中透露亘古而久远的芬芳。

This is a space with four bedrooms and two living rooms. The whole color tone leaves people with bright and expansive expressions. The designer applies neo-classical style for the decoration and carries out complex decoration with simple elements, performing delicate description from whole to detail parts. The interior design maintains the application of neo-classical style on materials and colors. From the space, one can intensively perceive traditional historical traits and profound cultural connotations. While at the same time, the design forsakes much too complicated texture and decorations, simplifying the lines. Here, both furniture and ornaments depict the elegant aristocratic status of the property owner with aesthetic posture and peaceful charms.

The whole interior space is based on light color tone, which makes the space bright and generous and produces open and generous temperament. Thus the space does not appear cramped at all. Furniture of dark wood color is accompanied with decorative ornaments of white or light wood colors, which is not depressing at all but visually produces some brand-new impressions. Furniture of color tone with primitive simplicity, natural materials and profound and implicit style display style advocating for naturalness. The format, texture, decoration and color tone of the furniture is refined and noble, revealing some everlasting fragrance from this enchantment.

富通天邑湾二期样板房

Futong Tianyiwan Phase 2 Show Flat

设计单位：深圳市昊泽空间设计有限公司
设 计 师：韩松
项目地点：广东省东莞市
项目面积：130 m^2
主要材料：白沙米黄大理石、白橡木、仿古镜、软包

Design Company: Shenzhen Haoze Space Design Co., Ltd.
Designer: Han Song
Project Location: Dongguan in Guangdong Province
Project Area: 130 m^2
Major Materials: Beige Marble, White Oak, Archaic Mirror, Soft Roll

室内空间的布置唯美、浪漫，用白色和浅米色搭配墙面，配上雅致的家具，营造出纯净高贵的美感。同时通过传统壁炉、时尚家具、柔美灯光、水晶材质等细节，演绎出精致与高雅。既有对历史的延续，又不拘泥于传统的思维逻辑，浅淡明快的色调、清晰的质感与肌理，营造了不一样的欧式时尚。

设计师在构思整体空间设计的时候，既把握了对历史的延续性，又不拘泥于传统的思维逻辑，将自己对新古典风格的理解用全新的方式演绎出来。浅淡明快的色调展现出设计师对色调的高超的把握能力，也勾勒出业主对完美生活的憧憬。不拘一格的设计手法，让居室在业主眼前焕然一新。配合着随处可见的清晰的质感与肌理，独具特色的欧式时尚就这样展现出来。

The arrangement of the interior space is aesthetic and romantic. The wall is decorated with white and light beige colors, accompanied with elegant furniture, producing some pure and elegant aesthetic beauty. Through details such as traditional furnace, fashionable furniture, soft lighting and crystal materials, the space appears delicate and elegant. With continuation of history, the space is not stuck in the traditional thinking logics. The light and brisk color tones and the clear texture create some different European fashion.

While carrying out space design conception for the whole space. The designer displays his understanding towards Neo-Classical style in some brand new way. The color tone presents the designer's supreme mastering of colors, while depicting the property owner's longing for perfect life. The unique design approach makes the space quite different. With clear texture everywhere, this spectacular European fashionable space is completed.

武汉复地东湖国际

Wuhan East Lake Intemationac Bay

设计单位：武汉王坤设计有限公司
设 计 师：王坤
项目地点：湖北省武汉市
项目面积：160 m²
主要材料：实木线条、木地板、石材、仿古砖、纸面石膏板、水曲柳饰面板、海基布、乳胶漆

Design Company: Wuhan Wang Kun Design Co., Ltd.
Designer: Wang Kun
Project Location: Wuhan in Hubei Province
Project Area: 160 m²
Major Materials: Solid Wood Lining, Wood Floor, Stone, Antique Brick, Paper Surface Plasterboard, Ashtree Veneer, Emulsion Paint

设计师为业主打造了一个高雅、素净的空间，从客厅到餐厅再转向书房，一路欣赏一路品位，设计师用不同材质和肌理演绎着多样的触觉和视觉表现，也让不同界面的相互关系有了精致且有趣的变化。慢生活的节奏在本案中被设计师展现到极致，客厅中舒适的沙发、窗边精致的靠背椅、拥有细腻纹样的地毯上散落的杂志，午后慵懒的阳光洒在实木地板上，带给人别样的温情。墙上黑白的摄影照片仿若沉静的时光，凝聚了那一刻的美丽并将它带入现在。简洁、洗练是设计的主要语言，同时设计师提倡比例均匀、材料搭配合理、收口方式干净利落、维护方便的原则，注重对细部的处理使得局部与整个空间的和谐一致。

在空间设计上，设计师最大程度让空间开放、通透、流动，整个空间动线明确，业主能够感受到自由、舒适的气氛。在家具与陈设的设计上，设计师辅以色彩和细部的变化，让整个空间充满了精致、典雅、唯美的气息。简洁的装饰线条让整个环境显得纯净而唯美、舒适而雅致，大气又不乏贵气。在这里，业主可以用心感受生活的真谛，从细节中发现美好。

The designer creates an elegant, pure and tidy space for the property owner. From the living room to the dining hall and the study, you can find designs of high taste. The designer makes use of different materials and texture to display diverse senses of touch and visual expressions, thus creating exquisite and funny variations for the mutual relationships of various interfaces. The rhythm of low pace life was displayed to extreme for this project, such as cozy sofa inside the living room, exquisite armchair beside the window and scattered magazines on the carpet with meticulous pattern. Lazy sunshine after noon spreads on the solid wood floor, creating some different warmth for people. The black and white picture on the wall is like some quiet time, condensing the beauty of that time and bring it to the present. Conciseness is the main language for design. While at the same time, the designer also advocates for the principles with well-distributed proportion, appropriate material collocation, clear closing technique and convenient maintenance. Emphasis on the treatment towards details make parts coexist in harmony with the whole space.

For space design, the designer makes the whole space open, transparent and fluent to the utmost. The whole space has clear lines, thus the property owner can feel some free and comforting atmosphere. For the design of furniture and layout, the designer applies the variation of colors and parts to make the whole space full of delicate, elegant and aesthetic atmosphere. Concise decorative lines make the whole space appear pure and nice, comforting and elegant, but no less aristocratic. Here, the property owner can attentively feel the essence of life, searching for beauties from details.

成都协信样板房

Chengdu Xiexin Show Flat

设计单位：重庆品辰设计装饰工程有限公司
设 计 师：庞飞
项目地点：四川省成都市
项目面积：200 m^2

Design Company: Chongqing Pinchen Decorative Design and Engineering Co., Ltd.
Leading Designer: Pang Fei
Project Location: Chengdu in Sichuan Province
Project Area: 200 m^2

本案以大气、优雅的欧式新古典风格为主，一进入客厅空间，顶棚上悬挂的奢华吊灯便彰显出了低调、奢华的气质，呼应着餐厅的水晶灯，晶莹透亮的灯具在客厅营造出贵气氛围，简洁、有秩序感的吊顶设计于大气中展现细腻。不同于客厅带给人的宫廷式的硬色调，工作室中浅色调的设定使空间从立面色彩到家具都减轻了压迫感。曼妙的流线形书桌赋予了空间一丝灵性，华美的服装、宛如水彩画一般艺术感十足的地毯，成为了空间中最浓墨重彩的一笔。

儿童房的设计充满童趣，玩偶和动漫画像随处可见。墙壁上几个颇具动感的剪影栩栩如生，为孩子带来了许多生活乐趣。起居室过道的设计也独具个性，采用时尚的不锈钢材质的隔断，镜面反射效果堪称完美，更为空间打造出了别具一格的现代风情，配合着弧形线条的沙发，让古典和时尚在此交融，演绎出使人眷恋的唯美时光。

This project is focused on magnificent and elegant European Neo-Classical style. Upon entering the living room, you would find the luxurious droplights hanging on the ceiling which display low-key and luxurious temperament, corresponding with the crystal lights in the dining room. The glittering and translucent lights create some noble atmosphere inside the living room. The concise and orderly ceiling presents some delicacy in the magnificence. Different from the hard color tone of palace style in the living room, the light color tone of the studio lessens the sense of tension of the space from elevation color towards furniture. The graceful streamline style table entrusts the space with some dynamic feel. The magnificent costume and the artistic painting like carpet become the spot thick and heavy in colors inside the space.

The design of the children's room is full of child's interests. You can find dolls and cartoons everywhere. The several dynamic sketches on the wall are true to life, bringing some life funs for the children. The design of the living room's corridor is also quite peculiar. The partition of fashionable stainless steel and the mirror surface with perfect reflective effect create some different modern fashion for the space. Accompanied with sofa of arch lining, classical style integrates with fashion here, producing some aesthetic time that people are sentimentally attached to.

清雅华尔兹

Graceful Waltz

设计单位：简艺东方设计机构
设 计 师：孙长健、林元娜
项目面积：295 m^2
摄 影 师：周跃东

Design Company: Simple Art Oriental Design Institution
Designer: Sun Changjian, Lin Yuanna
Project Area: 295 m^2
Photographer: Zhou Yuedong

本案最大的亮点在于其平面布局设计，这是一套前后带花园的叠拼别墅。原本有一面的花园连接着卫生间及保姆间，这样的设计抑制了花园的魅力与光彩。于是，设计师在布局上做了大幅度地调整，将卫生间与保姆间移开，在此设置餐厅，做成了一个带花园的宽大明亮的大餐厅，同时将厨房的外阳台也纳入其中，整合成一个大厨房。

原本二楼有两间卫生间，但面积都相对较小，在使用上不方便。因此，设计师在布局上进行了调整——取消起居室，将主卧设计成带更衣间的大套间，再保留一个卫生并将其扩大。开敞式的书房设计令人心旷神怡，更在空间上使二层与一层相互呼应，扩大了空间的视野，同时改善了二层的采光与通风。

The bright spot of this project lies in the plane layout design. This is a townhouse with gardens at front and in the back. The garden connects the washroom and the nurse's room. This design restrains the glamour and charms of the garden. Thus the designer makes some grand transformations towards the layout, removes the washroom and the nurse's room and set a dining hall here, thus creating a broad and bright dining hall with a garden. While at the same time, the outer balcony of the kitchen is included inside, producing a grand kitchen.

There are originally two washrooms on the second floor, but with comparatively small areas, which are not that convenient in practical use. Thus, the designer makes some adjustments towards the layout. The master bedroom is designed into a bid suite with dressing room by removing the living room, while maintaining a washroom and expanding it. The open study design makes people refreshed and relaxed, while making the second floor and first floor correspond with each other, expanding the views of the space and improving lighting and ventilation of the second floor.

荷塘月色

Moonlight over the Lotus Pond

设计单位：北京东易日盛长沙分公司
设 计 师：王帅
项目地点：湖南省长沙市岳麓山公馆
项目面积：320 m²

Design Company: Changsha Branch of Dongyirisheng Decoration
Designer: Wang Shuai
Project Location: Yuelu Mountain Mansion, Changsha in Hunan Province
Project Area: 320 m²

岳麓山公馆坐落在绿树成荫的河西岳麓山脚下，建筑外立面由仿古条形砖砌筑而成，带有北欧建筑特色的板岩屋顶散发着浓郁的乡村气息。住宅的前面建有鱼池，水系围绕整个小区流入池塘。在盛夏傍晚，可以坐在住宅前面的门廊上，感受着徐徐的微风，嗅着空气中传来淡淡的花香，聆听流水的潺潺声和此起彼伏的蛙鸣虫叫声，欣赏着满天星斗，依附在父亲的臂弯里，陶醉在父亲讲述的童话故事中。幸福的生活是如此的简单而又美妙动人。

男业主说："在全世界都在建造的大型商业住宅中，期望这栋住宅能够与众不同。"因此，他对本案有着非常高的标准。女业主也是位很有品位的女性，和蔼可亲、热情大方。业主与设计师在整个项目的实施过程中合作得非常愉快，直到今天，他们依然继续着这段友谊。

This project is located at the foot of Yuelu Mountain in Changsha's Hexi with shading trees. The building's exterior is made of antique lath bricks. The slate roofing has north European architectural characteristics, sending out some intensive rural atmosphere. There is a fish pond at the front of the mansion. The whole residence is surrounded with water systems which lead into the pond. At the dusk of a hot summer, you can sit on the the porch at the front of the mansion, enjoying the gentle breeze, fragrance of flowers in the air, murmuring of flowing water and the croaking of frogs. Under the starry night, a child would lean into the arms of his father and listen to the fairytale about the goddess in the moon told by the father. Happy life can be so simple and so marvelous.

The host said, "All across the world, people are building grand commercial buildings, and I want a residence completely different from that." Thus, he has set quite high standards for this project. The hostess is also a woman with fine tastes, amiable, passionate and generous. The property owner has a very good cooperation with the designer during the process of this project. Till today, they even continue that friendship established at that time.

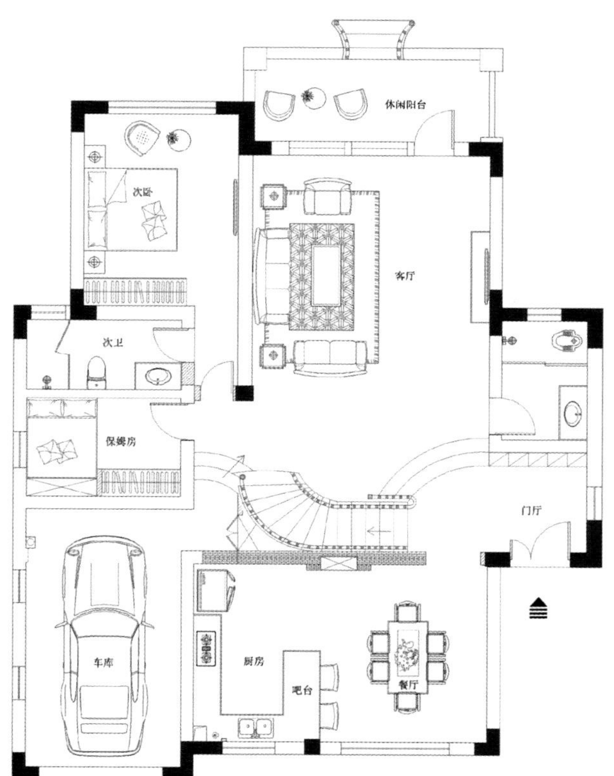

休闲阳台
次卧
客厅
次卫
保姆房
门厅
车库
厨房
吧台
餐厅

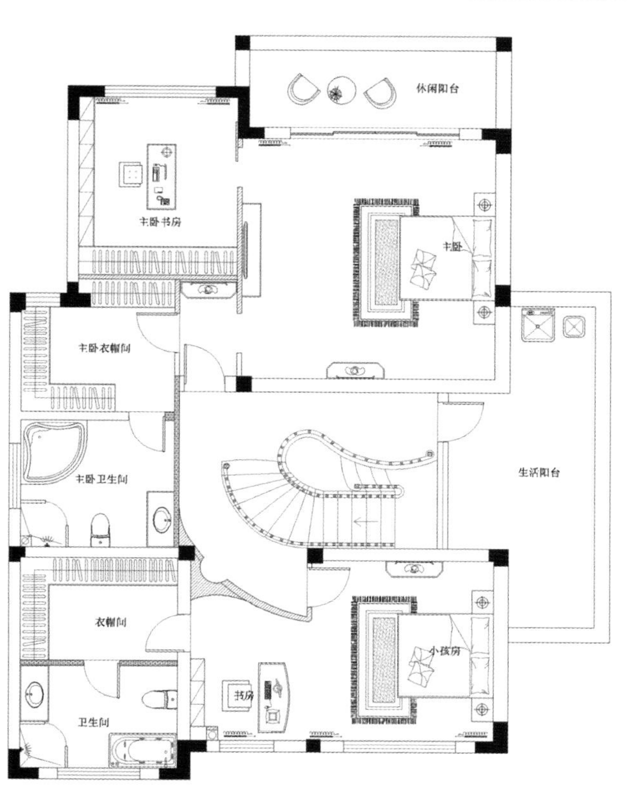

休闲阳台
主卧书房
主卧
主卧衣帽间
主卧卫生间
生活阳台
衣帽间
小孩房
书房
卫生间

格凌兰2期示范单元

Greenlem Phase 2, Show Flat

设计单位：之境室内设计事务所
设 计 师：廖志强
项目地点：四川省成都市
项目面积：120 m^2
摄 影 师：季 光

Design Company: Zhijing Interior Design Firm
Designer: Liao Zhiqiang
Project Location: Chengdu in Sichuan Province
Project Area: 120 m^2
Photographer: Ji Guang

ACCESSORIES
ACCESSORIES

精致奢华生活

本案是给豪运地产打造的 2 期示范单元，是在完成 1 期的示范之后的又一套精致示范单元。设计初期，对项目的整体定位进行了细致的分析，使本套示范单元以精致生活为主题。

户型上做了比较大的调整，给出了开放的空间概念。开放的入户厨房和开放的书房除了在空间上给人较大的视觉冲击，还给了空间很大的延展性，这也是置业公司一直想表达的。在完成开放概念的同时，并没有牺牲原本的生活所必须的刚性需求，同时还实现了主卧室衣帽间和独立卫生间的设置，使主卧室的生活品质得到了较大的提升。

Delicate Luxury Life

This project is another boutique show flat for phase 2 after the show flat of phase 1. For the initial stage of the project, the designer carries out meticulous analysis for the whole orientation, with boutique life as the theme for the show flat.

The designer makes some grand adjustment on the flat type, thus producing open space concept. Open indoor kitchen and open study not only produce some huge visual shocks for people, but also creates some grand extensive capability. And that is also what the properties corporation tries to express. While completing open concept, the design does not sacrifice the necessary rigid demand for life, but completes the setting of the master bedroom cloakroom and independent washroom, thus greatly promoting the life quality of the master bedroom.

锦园禧瑞·适度经典

Jinyuan Xirui•Appropriate Classical

设计单位：珠海空间印象建筑装饰设计有限公司
设 计 师：霍承显
软装设计：潘敏意
项目地点：广东省珠海市
项目面积：177 m^2
主要材料：白玫瑰大理石、水晶玫瑰大理石、土耳其灰大理石、中东米黄大理石

Design Company: Zhuhai Space Impression Architectural Decorative Design Co., Ltd.
Designer: Huo Chengxian
Soft Decoration Designer: Pan Minyi
Project Location: Zhuhai in Guangdong Province
Project Area: 177 m^2
Major Materials: White Rose Marble, Crystal Rose Marble, Turkey, Grey Marble, Middle East Beige Marble,

本案的设计采用舒适优越的经典风格，软装设计简约圆润，在配色之中添加少许活泼之感，使得家居在沉稳大气之余，整体散发出一种精致怡人的气度，适合追求完美居住环境的人群。工作之余回归家庭，能够将身心疲惫完全释放，舒展于此。

质感以陶瓷、布艺、木饰面作主打，色调温暖光洁，充满亲切感。客厅的米色沙发、绿白镶嵌的布艺地毯、镂空造型的木质屏风、陶瓷质地的花瓶摆件，看似简单的搭配，却将陶瓷、布艺和木饰面紧密结合，散发出各自的特质之余摩擦出新的气质。卧室以米色为主调，以陶瓷布艺增添灵动之感，温馨舒适之中隐现生活的巧妙活泼。

This project has comforting and superior classical style, with concise and mellow soft decorative design. The designer adds some dynamic colors to make the whole space send out some pleasing temperament based on the sedate and magnificent atmosphere. This space is appropriate for people seeking for perfect residential environment. After you get home from work, you would get complete relaxation here.

The texture focuses on ceramics, cloth and wood veneer. The color tone is warm and clear, full of intimacy. Inside the living room, there is beige sofa, green and white cloth carpet, wood screen of hollow-out format, ceramic vase ornament, etc. The seemingly simple collocations closely connect the ceramic, cloth and wood veneer, creating some new temperament while displaying the peculiar temperament. The bedroom has beige as the tone and uses ceramics cloth to display some dynamic sensation. There is some lively dynamic and vivid sphere among this warmth and comfort.

請勿使用

格林兰景

Greenland

设计单位：福州筑易装饰设计有限公司
设 计 师：郭娟
项目地点：福建省福州市
项目面积：115 m²
主要材料：乳胶漆、瓷砖、艺术壁纸、镂空雪弗板、大理石

Design Company: Fuzhou Zhuyi Decorative Design Co., Ltd.
Designer: Guo Juan
Project Location: Fuzhou in Fujian Province
Project Area: 115 m²
Major Materials: Emulsion Paint, Ceramic Tile, Artistic Wallpaper, PVC Expansion Sheet, Marble

KELLY HOPPEN HOME

ROOMS TO INSPIRE
KELLY HOPPEN HOME

印象中，欧式的家居空间总是与金碧辉煌的奢华感密不可分。而本案设计师却反其道而行之，摒弃了张扬的奢华装饰，不显山不露水地为业主呈现出一种别样的欧式家居空间。触目所及为温暖的橡木色，纹理清晰、雕刻精细的家具，绽放出的不仅仅是一种岁月沉淀之下的宁静，更是一种静水长流的低调的华贵感。

在整个空间里，设计师巧妙运用各种品位独到的元素，如同经典交响乐中的悠扬篇章，在隽永和内敛中，尽情展现居住者的优越与不同凡响。在景观阳台和厨房的墙壁上，设计师匠心独具地运用错落的线条勾勒出不规则的空间跳跃感，使整个空间充满了强烈的现代气息，凸显出本方案的个性。

In people's impression, European residential space is always closely interconnected with the resplendent luxury feel. But the designer of this project goes totally against it, forsaking the showy and luxurious decorations and displaying some peculiar European residential space for the property owner in some subtle way. All in eyes are warm oak color, the furniture of clear pattern and refined carving sends out not only some tranquility of the time, but also some low-key luxury feel of everlasting charms.

Inside the whole space, the designer ingeniously makes use of some elements of peculiar tastes, which are like the melodious chapters in the classical symphony, displaying the superiority and outstanding quality of the property owners to heart's contents. On the landscape balcony and the kitchen wall, the designer skillfully makes use of lines strewn at random to depict the bouncing feel of the irregular space, making the whole space full of intensive modern atmosphere, and highlighting the characteristics of this project.

大儒世家·绿园

Daru Shijia•Green Garden

设计单位：私享空间设计事务所
设 计 师：真志松
项目地点：福建省福州市
项目面积：82 m^2
主要材料：大理石、软包、白镜

Design Company: Sixiang Space Design Studio
Designer: Zhen Zhisong
Project Location: Fuzhou in Fujian Province
Project Area: 82 m^2
Major Materials: Marble, Soft Roll, White Mirror

本案以简洁、大气的白色为主调，以打造优雅、明快的欧式新古典风格为主题。现代社会快节奏的生活方式，使年轻人在选择生活空间时，更加倾向于能够舒缓身心、充分贴合自身需求的设计。因此，简洁、实用、舒适是本案设计的基本原则，业主与设计师沟通时，也希望在经济、实用、舒适的同时，能体现出一定的品位。所以，本案的设计不仅注重居室的实用性，而且还体现出现代社会生活的精致与个性，符合业主的生活品位。经过设计师全面的考虑，在总体布局方面尽量满足业主生活上的需求，在外观上呈现精致简约的质感，主要装修材料以爵士白大理石、皮革软包、白镜面为主。

客厅是家居生活的重要空间，也代表着业主的品位，体现出业主优良的品格及社会地位，也是业主与朋友交流、娱乐的重要场所。设计师在电视背景上运用爵士白大理石与米黄色皮革软包相结合的设计手法，既简单又大方，沙发背景的白镜采用电脑车花工艺，不仅延伸了空间的美感，还能体现时尚的品位。

This project has concise and magnificent white as the tone color and elegant and brisk neo-classical European style as the theme. Due to the fast pace life style of modern society, while selecting life spaces, young people tend to select designs which can soothe their heart and fit their personal requirements better. Thus, the basic principles of this project focus on being concise, practical and comfortable. While the property owner negotiates with the designer, he also hopes that the space can display his taste while maintaining economic, practical and comfort requirements. Thus, the design of this project not only emphasizes on practicality of the interior space, but also reflects the delicacy and individuality of modern social life, in accordance with the life taste of the property owner. Through the comprehensive considerations of the designer, the overall layout should meet with the property owner's life requirements at most and the outlook should display exquisite and concise texture. The major decorative materials focus on jazz white marble, leather soft roll and white mirror surface, etc.

Living room is an important space for family life, which represents the taste of the property owner, his fine characteristics and his social status. And that is also an important location for the property owner to negotiate and entertain with his friends. For the TV background wall, the designer applies the design approach with the combination of jazz white marble and beige soft roll, which is simple but magnificent. The white mirror applies computer embroidery technique which not only extends the space aesthetic beauty, but also represents fashionable taste.

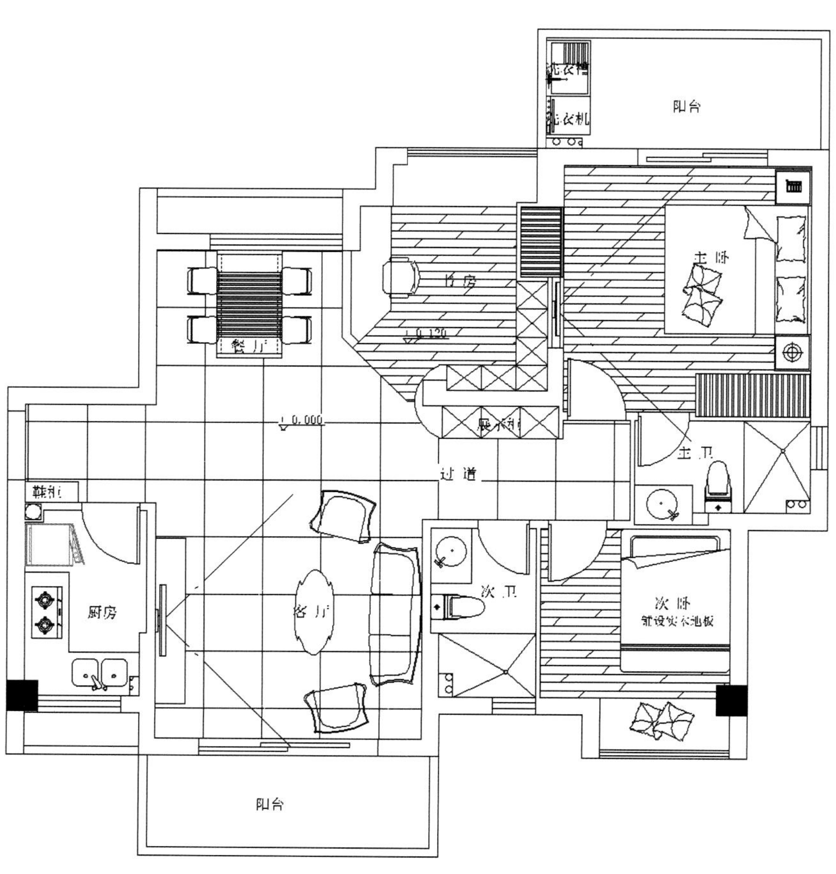
阳台
主卧
书房
餐厅
0.000
主卫
过道
厨房
客厅
次卫
次卧
阳台

中源名都21号楼三层B2户型

Zhongyuan Mingdu, Flat Type B2, 3rd Floor, Building No. 21

设计单位：J2–DESIGN/ 厚华装饰设计有限公司
设 计 师：卓晓初
陈设与选材师：谭启开
项目地点：广东省肇庆市
项目面积：97 m^2
主要材料：银镜、黑镜、地毯、软包
开 发 商：广东新中源集团

Design Company: J2-DESIGN/Houhua Consultant Design Co.,Ltd
Designer: Zhuo Xiaochu
Layout and Material Deisnger: Tan Qikai
Project Location: Zhaoqing in Guangdong Province
Project Area: 97 m^2
Major Materials: Silver Mirror, Black Mirror, Carpet, Soft Roll
Developer: New Zhongyuan Group in Guangdong Province

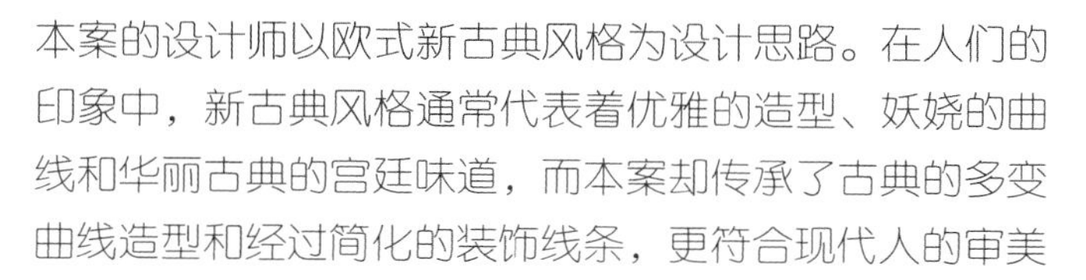

本案的设计师以欧式新古典风格为设计思路。在人们的印象中，新古典风格通常代表着优雅的造型、妖娆的曲线和华丽古典的宫廷味道，而本案却传承了古典的多变曲线造型和经过简化的装饰线条，更符合现代人的审美情趣，同时依然保持着古典的贵族气息。居所整体以暖色调为主，搭配蓝色、黑色作为点缀，传递出生活中的唯美和温馨。客厅采用六边形蜂窝边框组合，无论在地面、立面、顶棚中都运用到该元素，上下的融合使其对空间起到画龙点睛的作用。

在整体样板房中设计师通过黑白的对比加强视觉冲击力，再以灰色调缓和局部空间的气氛，配合精致典雅的装饰物，给人舒适、自然之感。客厅中用白色鸟笼的造型来装饰绿植花卉，独特的造型给空间带来些许趣味性，更重要的是让空间融入了自然的气息。卧室采用竖纹壁纸与竖条软包来装饰空间，淡雅的蓝色壁纸使卧室气氛安静而又平和，饰有华美纹样的窗帘给卧室平添了几分唯美的气息。

从起居室前方的过道缓步而行，即进入专属于业主的私密空间。从起居室另一侧的拉门穿越后方过道，则进入次卧与另一卧室的私密空间。透过不同过道入口的分隔，界定了主卧的私人独享空间。此外，经由过道的缓步进入，不但沉淀了心情，也加深了对私密空间的期待。金、银、白、紫等带有金属光泽质感的色彩，画框裱布、编织壁纸、拼花黑檀墙面、银箔造型电视墙、玻璃马赛克、贴饰孔雀壁纸——设计师在本户的私密空间中，大胆运用了多元且绮丽的材质，堪称“风华园”中豪门风格的代表作。同时，也将 Art Deco 装饰主义的艺术美学发挥得淋漓尽致，令人震撼而惊艳。

The designer of this project takes neo-classical European style as the design train of thoughts. In people's eyes, neo-classical style usually represents elegant posture, enchanting curves and magnificent palace taste, better appropriate for modern people's aesthetic interests while at the same time maintaining its classical noble atmosphere. The whole residence has warm color tone, accompanied with blue and black as the decorative ornaments, which sends out the aesthetic and warmth of life. The living room applies the combination of hexagon honeycomb frame. The ground, elevation and ceiling all apply this kind of elements. The integration of upper and lower spaces makes the finishing point for the space.

Inside the show flat, the designer strengthens the visual impact through the contrast between black and white. The designer uses grey color to soften the atmosphere of part of the space and creates some comforting and natural atmosphere through the accommodation of delicate and exquisite decorative objects. Inside the living room, the white bird cage decorates the green plants and flowers, bringing some interests for the space with its peculiar formats. What's more important, the design produces some natural atmosphere for the space. The bedroom is decorated with wallpaper of vertical pattern and vertical soft rolls. Elegant blue wallpaper makes the bedroom atmosphere serene and peaceful. The curtain of magnificent pattern produces some aesthetic atmosphere for the bedroom.

You can walk slowly on the corridor at the front of the living room and come into the private space exclusive to the property owner. From the back corridor through the sliding door on the other side of the living room, you can come to the secondary bedroom and the private space of the other bedroom. Through the partition of different corridor entrance, the private exclusive space of the master bedroom is defined. Other than that, while you walk slowly along the corridor, you are calm down and get further expectation towards the private space. The colors of metallic luster, such as gold, silver, white and purple, frame cloth, knitted wallpaper, ebony wall, TV background wall of silver foil format, glass mosaic tile, decorative peacock wallpaper... For the private space of this project, the designer boldly applies various but splendid materials to create a masterpiece of royal family style. While at the same time, artistic aesthetic beauty of Art Deco style is displayed to the full, quite impressive and amazing.

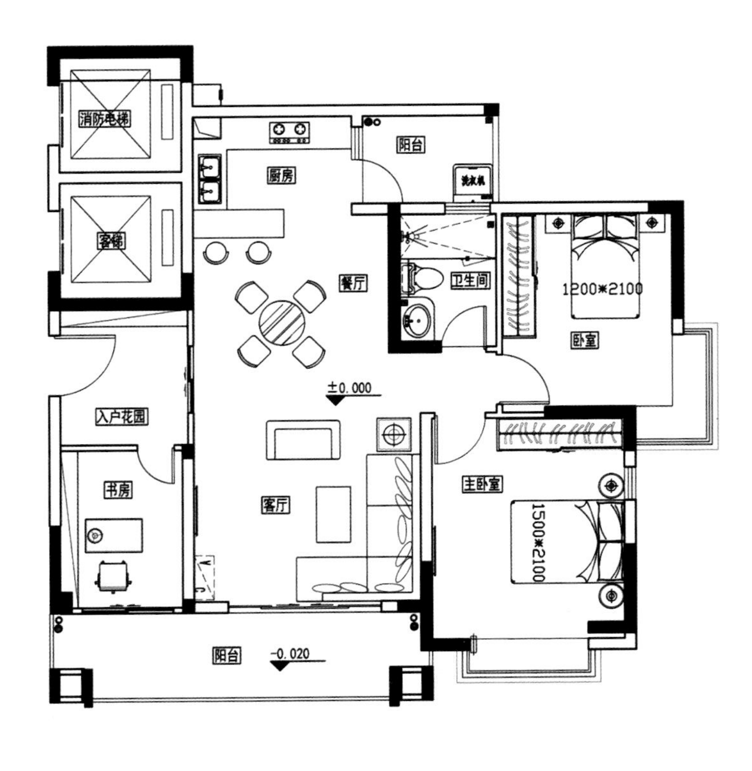
消防电梯
客梯
厨房
阳台
洗衣机
餐厅
卫生间
1200*2100
卧室
±0.000
入户花园
书房
客厅
主卧室
1500*2100
阳台
-0.020

中南世纪城
Zhongnan Century City

设计单位：TY34 精品室内设计中心
主设计师：庄光科
参与设计（包含软装设计）：马翠琴、杨连文
项目地点：江苏省镇江市
项目面积：160 m^2
主要材料：抛光砖、爵士白大理石、
啡网纹大理石、镜面玻璃
摄 影 师：金啸文

Design Company: TY34 Boutique Interior Design Center
Chief Designer: Zhuang Guangke
Associate Designers: Ma Cuiqin, Yang Lianwen
Project Location: Zhenjiang in Jiangsu Province
Project Area: 160 m^2
Photographer: Jin Xiaowen
Major Materials: Polished Brick, Jazz White Marble, Cobwebbing Marble, Mirror Surface Glass

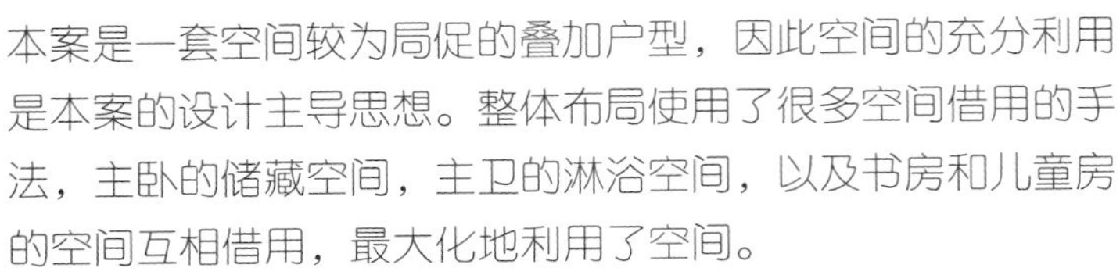

本案是一套空间较为局促的叠加户型，因此空间的充分利用是本案的设计主导思想。整体布局使用了很多空间借用的手法，主卧的储藏空间，主卫的淋浴空间，以及书房和儿童房的空间互相借用，最大化地利用了空间。

整套设计运用了一些后现代的设计手法，用高反光面的材质来表现空间的质感。纯白色突出了整体空间的扩张感，让空间显得更加唯美。在线角的把握上，摒弃了很多欧式复杂的做法，用两层退阶的做法代替了欧式繁复的线条，整体空间既简约又奢华。

家具全部为定制的纯白色的家具，以欧式家具为蓝本，做了一些局部的修改，保留了原有家具的奢华，又在色调和款式上更加简洁；陈设多为镀铬的欧式摆件、烛台，布艺均为银色和白色搭配的样式，与整体风格相统一；灯具主要是玻璃弯管和白色烤漆的风格互相穿插，每个区域的壁灯、吊灯、台灯都互不干扰，相得益彰。

This project is a multiplied flat type with cramped space. Thus it is the leading philosophy of the design to make full use of the space. The whole layout makes use of many space borrowing approaches. The storage space of the master bedroom, the shower space of the master washroom and the mutual application of study and children's room maximize the use of the space.

The whole design makes use of some post-modern design approaches and displays the texture of the space with some materials with some high reflection of light. Pure white color highlights the expansion feel of the whole space, which makes the space much more aesthetic. As for the aspect of lines and corners, the designer forsakes many complicated European practices. The designer replaces complex European lines with two layers, thus making the whole space concise and luxurious.

All the furniture are custom-made pure white, with European furniture as chief of source, making some changes on the parts and maintaining the luxury of the original furniture, and being more concise on color tones and fashions. The layout is mostly chrome plated European accessories and candle sticks. The cloth is mostly pattern of the combination of silver and white colors, consistent with the whole style. Lights mainly center on style of glass bending tubes and white baking finish. Wall lamps, drop lights and desk lamps of every area are independent from each other, bringing out the best in each other.

中源名都21号楼A户型

Zhongyuan Mingdu, Building No. 21, Flat Type A

设计单位：J2–DESIGN/ 厚华装饰设计有限公司
设 计 师：黄炽烽、欧敏华
陈设与选材：区婷婷、徐流艳
项目地点：广东省肇庆市
项目面积：150 m^2
开 发 商：新中源地产
主要材料：木饰面板、绒面布、抛光砖

Design Company: J2-DESIGN/Houhua Consultant Design Co.,Ltd
Designers: Huang Chifeng, Ou Minhua
Layout and Material Supervisor: Qu Tingting, Xu Liuyan
Project Location: Zhaoqing in Guangdong Province
Project Area: 150 m^2
Developer:New Zhongyuan Estate
Major Materials: Wood Veneer, Velvet Cloth, Polished Tile

本案从营造都市艺术情调入手，设计师通过多种材质的搭配打造出神秘的都市氛围。在平面布局上，设计师设置了开放式的厨房，使其连通餐厅和入户花园，从而使空间在使用上更加灵活。在墙面材料上，设计师用少量材料如木饰面和绒面布硬包的搭配，彰显出空间的奢华、贵气，让空间散发出独特的味道。地面采用多种拼花设计，不仅有效地界定出各个区域的范围，更迎合了整个空间的装饰效果，成为空间的主题之一。

软装设计中，主体墙上的装饰品是塑造空间质感的重要元素。充满艺术感的金属装饰图案赋予了客厅独特的主题和个性，配合着华美的红色绒面沙发以及深色木质茶几上的抽象线条的摆件，给人一种奢华、高贵的感受。

This project starts from creating urban artistic emotional appeal. The designer creates mysterious metropolitan atmosphere with the combination of various materials. For plane layout, the designer sets open kitchen which connects with dining hall and indoor garden. Thus the use of space is much more flexible. As for wall materials, the designer presents the luxury and aristocracy of the space with some materials such as the combination of wood veneer and velvet cloth hard roll, making the space send out some peculiar taste. The ground applies multiple pattern design which not only effectively demarcates the scope of every region, but also caters to the decorative effects of the whole space, becoming one theme of the space.

As for soft decoration design, the decorative object on the main wall is an important element in shaping the texture of the space. The metal decorative objects with artistic feel give the living hall peculiar themes and characteristics. Accompanied with magnificent red velvet sofa and goods of furniture for display with abstract lining on the dark color wood tea table, people are impressed with this luxury and nobility.

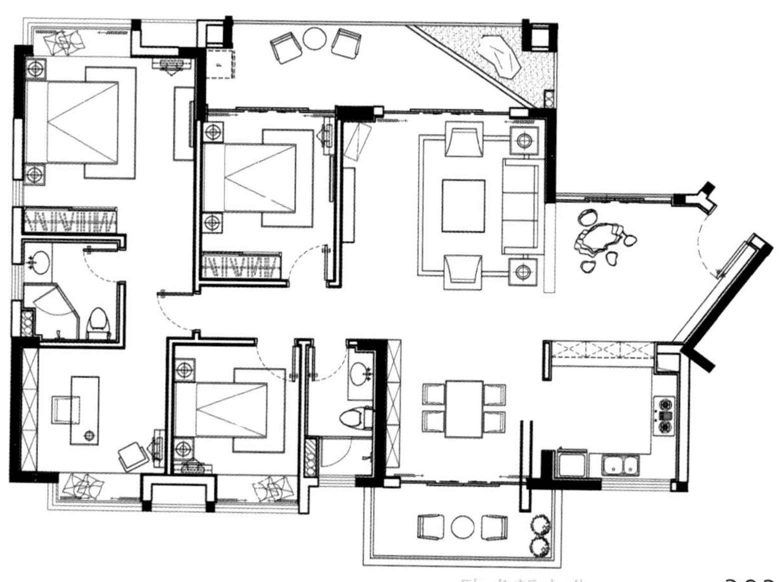

图书在版编目(CIP)数据

摩登样板间.2,欧式新古典/ ID Book工作室 编 —武汉：华中科技大学出版社，2013.6
ISBN 978-7-5609-8781-1
Ⅰ. ①摩… Ⅱ. ①I… Ⅲ. ①住宅－室内装饰设计－图集 Ⅳ. ①TU241-64

中国版本图书馆CIP数据核字(2013)第056631号

摩登样板间II·欧式新古典

ID Book工作室 编

出版发行：华中科技大学出版社（中国·武汉）
地　　址：武汉市武昌珞喻路1037号（邮编：430074）
出 版 人：阮海洪

责任编辑：曾　晟
责任校对：王孟欣
责任监印：秦　英
装帧设计：吴亚兰

印　　刷：天津市光明印务有限公司
开　　本：965 mm×1270 mm　1/16
印　　张：19
字　　数：275千字
版　　次：2013年6月第1版第1次印刷
定　　价：328.00元(USD 69.99)

投稿热线：(010)64155588-8000 hzjztg@163.com
本书若有印装质量问题，请向出版社营销中心调换
全国免费服务热线：400-6679-118 竭诚为您服务